Understanding Local Economic Development

Understanding Local Economic Development

EMIL E. MALIZIA

EDWARD J. FESER

CENTER FOR URBAN POLICY RESEARCH
Rutgers, The State University of New Jersey
New Brunswick, New Jersey

Third Printing 2005

Published by the **CENTER FOR URBAN POLICY RESEARCH**
Edward J. Bloustein School of Planning and Public Policy
Rutgers, The State University of New Jersey
Civic Square • 33 Livingston Avenue • Suite 400
New Brunswick, New Jersey 08901–1982

Printed in the United States of America

Library of Congress Cataloging-in-Publication Data

Malizia, Emil E.
Understanding local economic development / Emil E. Malizia and Edward J. Feser.
p. cm.
Includes bibliographical references and index.

ISBN 0-88285-163-2 (alk. paper)

1. Economic development. 2. Economic policy. 3. Regional development—United States. 4. Local government—United States. 5. United States—Economic policy—1993– I. Feser, Edward J. II. Title.
HD75.M25 1999
338.9—dc21 98-44999
CIP

Cover design: Helene Berinsky
Cover photograph: U.S. Department of Agriculture

For

Dorothy and Emilio Malizia

and

Charles Draves

Contents

Figures and Tables

Figures

Tables

Preface

Economic development encompasses a wide range of concerns. To most economists, economic development is an issue of more economic growth. To many business leaders, economic development simply involves the wise application of public policy that will increase U.S. competitiveness. To those who think that government should more actively direct the economy, economic development is a code phrase for industrial policy. To environmentalists, economic development should be sustainable development that harmonizes natural and social systems. To labor leaders, it is a vehicle for increasing wages, benefits, basic education, and worker training. To community-based leaders and professionals, economic development is a way to strengthen inner-city and rural economies in order to reduce poverty and inequality. To public officials at state and local levels, economic development embodies the range of job-creation programs broadened since the 1980s in response to the decline of federal domestic assistance.

Economic development as practiced at the local level is a technique-dominated field concerned with increasing jobs and the tax base, primarily by marketing the location to prospective and existing employers. The political culture in which most developers operate emphasizes short-term, quick-fix solutions, as well as the emulation of development strategies and programs in successful communities. Thus, the practice of economic development is strikingly similar across the United States as economic developers try to keep up with the competition. Like bankers offering a standard commodity (money) at slightly different terms and rates, economic developers endeavor to show that their location is better than any others on a prospect's short list and that their incentive package is as good as any offered by the competition. Yet, unlike the money bankers rent, localities differ greatly, as is aptly demonstrated by differences among their firms, labor forces, politics, natural resources, and geography.

This book has an ambitious objective. It is written to give current and future economic developers and community leaders in the United States knowledge they can use to understand both the process and the practice of local economic development. With such knowledge, they should have the confidence to think strategically about their community and to design *unique* ways to build on its strengths and overcome its weaknesses. *Understanding Local Economic Development* focuses on theory because information is a source of power, and well-constructed theory is the most powerful form of information. Collectively, the theories covered in this book contain the basic concepts developers need to know to analyze their local economies, define economic development, design effective strategies, evaluate the outcomes of different development activities, and communicate successfully with stakeholders whose support is needed to take action. Without theory, the developer can do only what is politically feasible. With theory, the developer has a logical tool for thinking about development, as well as an independent basis on which to build the local consensus needed for effective action. In a world cluttered with sound bytes and biased by spin, theory offers a safe harbor for reflection on the development process and, if politics allow, a foundation for rational action undertaken in the best interest of the general public.

Economic development theories abound. Varying in basic, fundamental ways, they make different behavioral assumptions, use different concepts and categories, explain the development process differently, and suggest different policies. The theories used by economic developers determine, either explicitly or implicitly, how these developers understand economic development, the questions they ask about the process, the information they collect to analyze development, and the development strategies they pursue. Ultimately, theoretical insights influence how successful economic developers are in promoting local competitiveness.

The organization of this book should help economic developers grasp the theoretical differences and select the most powerful theories for addressing the economic and political realities they face. The book is divided into two parts. Part I, "Fundamentals," presents an historical sketch of U.S. development practice, as well as the fundamental definitions and concepts needed to understand economic development theory. In addition, Part I contains two overviews, one that provides a frame of reference with which to grasp and apply each theory, and another that sketches the major schools of economic thought. In Part II, "Theories of Economic Development," nine chapters systematically present key, relatively self-contained theories of the local economic growth and development process. A synthesis is offered in the final chapter.

The systematic treatment of theory and its application to the U.S. context are important distinguishing features of this book. Most existing books on economic development theory are concerned with less-developed countries; they are often collections of seminal articles or combine presentation of theory with other objectives, such as analysis of regional policies or discussion of technique. We believe that there is little need to justify the utility of a U.S. focus from a practitioner's perspective, even while admitting the benefits one gains from studying international examples. Although combining theory and policy analysis can be useful, doing so often necessitates a brief theoretical component. Also, collections of articles, though possessing the advantage of bringing together in one book experts on particular theories and perspectives, often lack a comprehensive, summarizing flavor. By way of its basic organization and synthesis, this book presents one view of how an extremely broad body of ideas fits together. Although the view is not beyond criticism, it has proved useful in more than twenty-five years of teaching the concepts of development to future practitioners.

This book, then, does not supplant, but rather supplements, other contributions. Important early edited volumes include Friedmann and Alonso (1964) and Perloff and Wingo (1968). More recently, contributing authors in Bingham and Mier (1993) summarize economic development theories drawn from various social sciences and apply these ideas to local practice. Several books examine development theories and policy in the United Kingdom and Europe, including Chisholm (1990) and Armstrong and Taylor (1978, 1985). Gore (1984) provides an excellent critical review of Anglo-American regional theory, while Blakely (1994) gives a brief summary of theory in his wide-ranging and elementary introduction to economic development planning. Among the books most comparable to the present one are Richardson (1973, 1978), Blair (1991), and Higgins and Savoie (1995). In textbook style, Richardson presents theories and concepts developed through the mid-1970s, while Blair (1991) emphasizes regional economics and its application to policy issues. Most recently, Higgins and Savoie (1995) provide a comprehensive survey of regional development theories, along with an analysis of regional policy in Canada and the United States. Although Canadian practice is not referenced here, Higgins and Savoie's analysis would suggest that development activity in the United States and Canada holds many similarities; thus, this book hopefully will also prove useful to Canadian development practitioners.

This book is aimed at four audiences: economic development professionals, undergraduate and graduate students of economic development, economic development educators and researchers, and community leaders involved in formulating economic development policies or funding economic development programs. Designed to span the divide between theory and local economic development practice, it presents, in systematic fashion, theories, models, concepts, and perspectives that should enable readers to understand the essential features of local economic development in the United States.

The book is organized to serve both as a text for in-service training and university courses and as a reference to what is a broad and disparate literature. We have assumed that most readers have limited time, minimal background in economics, and little patience for jargon. In developing the organizational framework, however, we assumed a reader who is seriously motivated to use theory to improve economic development practice. Each chapter of Part II consists of three sections: (1) a presentation of the basic tenets of the theory; (2) a summary of typical applications of the theory as well as, in some cases, a direct application to a specific local economy; and (3) a more detailed discussion that alternately elaborates on the theory and presents major critiques. We expect that many development practitioners will initially want to read the first two sections of these chapters, the first section to gain a basic understanding of the theory, the second to gauge its practical usefulness. Practitioner-readers may then refer to the more detailed discussion section, as interested, for additional information and elaboration. Students of economic development will want to read the chapters in their entirety, while also paying attention to the appendices, which provide more abstract or background material. Throughout the book, discussion questions are provided to review and extend important ideas. The extensive endnotes offer greater depth and theoretical richness, which scholars should find useful.

Acknowledgments

This book reflects our joint teaching, uncounted discussions, and multiple rounds of writing and exchanged drafts. In this sense, identifying our separate contributions is a hopeless exercise. On the other hand, the history of the book and its own trajectory of development in terms of our rough division of labor may help place it in context. *Understanding Local Economic Development* had its genesis in a set of lecture notes developed by Emil Malizia over twenty-five years of teaching regional economic development theory to students in the Department of City and Regional Planning at the University of North Carolina at Chapel Hill. Although the book is substantially changed since the notes were first mimeographed for use in class, the organizational framework, and particularly the treatment of fundamentals and syntheses of theories in Part I and Chapter 11, strongly reflect Malizia's thinking about development as it evolved through interaction with students and colleagues. In addition to providing further development and refinement of the framework and concepts of Part I, Ed Feser took primary responsibility for several chapters in Part II. He also focused on introducing more recent concepts and models of development. We hope that our joint efforts, as two people who were trained and began forming their ideas of economic development during different social, economic, and academic periods (Malizia in the 1960s, Feser in the 1990s), lend a richness to the book that would otherwise not be possible. Every chapter in the book bears, to one degree or another, the imprint of each of us.

We would like to thank the many economic development students and professionals with whom we have worked. They have helped us identify the truly important concepts of development and learn how to interrelate theory and practice. Reading the regional theory literature has led us to appreciate the importance of historical thinking and the continued relevance of books and articles written before 1960. We hope that this respect and appreciation for important early work comes through clearly. Faculty colleagues at the University of North Carolina and at other institutions offered many insights through conversations and written work over the years. We acknowledge, in particular, the very useful comments from the editor in chief of CUPR Press, Robert Lake, and chapter-specific reviews from John Accordino, Nancy Ettlinger, Ed Malecki, Ernest Sternberg, and Rebecca Winders. Support through DCRP from Ed Kaiser facilitated timely completion of the work, as did help from Steve West and Carolyn Turner, who provided reviews and production assistance, respectively. Karen Becker helped produce the final manuscript and generated the figures and final tables.

PART I

Fundamentals

1

The Practice of Economic Development

Economic development as a practice has come of age. During the 1980s, the economic expansion of the global economy, the collapse of Soviet Communism, and successive Republican administrations in the United States combined to underscore the power of market economies to generate growth and create jobs. The 1980s boom supported the idea of less government regulation and more privatization of public services. Proponents of supply-side economics and entrepreneurial business development claimed economic growth and job creation as the outcome of their economic policies, despite evidence that Keynesian-style deficit spending and corporate restructuring were the more proximate causes. In the 1980s, the competitiveness of the U.S. economy within the global system became the central concern of economic development.

In the 1990s, the economy recovered from a brief recession and resumed its growth. Job creation and new business formations have occurred at an impressive rate. Equity markets continue to achieve historic highs, while the federal government continues to retrench as it struggles to bring spending into line with revenues. Statistical evidence shows increasing inequality in the distribution of real household income, which reflects the growing geographic separation of different social groups, both within urban areas and between regions. The optimism of the 1980s has thus given way to the sober recognition of persistent economic problems brought into relief by media images of poverty, homelessness, and inner-city decline. These images emphasize the continued importance of distributional issues as part of the economic development agenda.

Local Practice

Economic developers in the United States promote the public interest by facilitating community development that supports business development. They are concerned primarily with local competitiveness, although distributional and environmental issues are increasingly being recognized as well. Developers want to know whether the local economy with which they are affiliated will experience sustainable growth as part of a global market system that grows more complex and competitive every day. To answer this question skillfully, and to implement the answers effectively, local developers must engage in careful analysis of the economy, creative synthesis of alternative strategies, and adroit execution of development programs and techniques.

Until fairly recently, the economic developer did not need such a broad range of skills. During the 1950s and 1960s, local developers, particularly those in the South, were "industrial developers" seeking economic growth through investment that would create jobs and expand the local tax base. The industrial developer wanted to sell the locality and its industrial sites to prospects seeking locations for manufacturing facilities. The developer used various analytical tools and techniques, not so much to understand the economic base as to advertise, market, and sell the location to industrial prospects. The developer established ties to existing industrialists and mobilized their assistance as volunteers in promotion and recruitment efforts. It was necessary to be an effective facilitator and, more important, to be a good salesperson (Levy 1990, 1992).

Industrial developers were often at odds with groups concerned with equity or environmental issues rather than growth. These groups served different constituencies, supported different interests, sought different objectives, and viewed the development process differently. Developers pursued the benefits of economic growth, while these local interests recognized—and tried to mitigate—the social costs of economic growth.

Although important differences in orientation remain, in many places the forces of economic change are bringing these groups together as associates, if not allies. Now local developers are "economic" developers. This label not only implies concern for the entire economy in addition to the manufacturing base, but the need for new analysis and creative problem-solving skills to back up marketing and salesmanship efforts. As noted, competitive pressures within the framework of the global market system make local competitiveness the central concern. But competitiveness is a function of the quality of local resources as well as their availability and cost. At the same time, the local economy, more than external state and federal funding, has become the primary source of financial support to address the concerns for community development and environmental quality. Local borrowing capacity for capital projects depends on the strength of the local economic base.

Increasing complexity requires that economic developers possess a better understanding of the local economy and the nature of its internal connections and its connections to the global economy. Theory helps developers grasp the evolving development process. It provides insights about the relative attractiveness of one place, currently and in the future, compared to other places, as well as about the historical process that has generated its current status.

A Brief History of U.S. Economic Development Practice

Before 1960, concern for economic development was largely confined to the developing nations of the "Third World."[1] The exception was the American South. Henry Grady's call in the 1890s for economic progress and Mississippi's bold Balance Agriculture with Industry Program in the 1930s prompted southern states to hire "industry hunters" and unleash them on corporate boardrooms in the North (Cobb 1993). John F. Kennedy's campaign for the presidential nomination raised the national awareness of urban and rural poverty in the United States. The Great Society programs that followed were designed to eliminate "pockets of poverty" in center-city ghettos, declining rural areas, and depressed regions. The Economic Development Administration (EDA) was established to attack area poverty and unemployment. Through planning and technical assistance, grants to localities, and loan programs, the EDA promoted local economic development in areas on a "worst-first" basis. The Appalachian Regional Commission (ARC) and other federal and regional agencies were given similar mandates to promote development in lagging areas.

Throughout the 1970s, the practice of economic development increasingly became a local activity. Attempts were made to forge links among federal employment, social service, small business development, and economic development programs to increase their local effectiveness given ever more pressing federal budgetary constraints. States and localities took concrete steps to address the related problems of economic adjustment and fiscal stress. Every state developed industrial recruitment and promotional programs to bid for the investments of U.S.-based and foreign corporations. Many localities, especially the large jurisdictions in mature industrial areas, had been encouraging job creation and economic revitalization for some time. To provide resources needed for community development, business and neighborhood organizations supported local economic growth and development. Every jurisdiction appeared to be concerned with local economic development whether its economic base was growing, declining, stagnant, or experiencing readjustment problems. Economic developers were hired to address these problems.

During the 1980s, concern for local economic development remained pervasive, but the approach changed dramatically. At the federal level, the main threat was no longer poverty or domestic social unrest but competitive pressures from the international economy. The goal shifted from elimination of regional disparities and urban and rural poverty to enhancement of productivity, economic growth, and global market share. At the state level, traditional industrial development approaches were viewed as insufficient, and new policies and programs to promote innovation and entrepreneurship, and support existing industries, were developed (Osborne 1988). At the local level, seeking federal grants and corporate facilities came to be considered a high-risk approach to economic development as interjurisdictional competition increased. With corporate investment patterns becoming less predictable and domestic assistance declining, developers tried to refine externally oriented techniques (for example, targeted industrial recruitment) while adding internally oriented techniques for job creation. They pursued the talent, resources, and ideas that could provide employment opportunities and shore up the tax base from any available source. Over time, as the basic tools of the trade proliferated, the size and scope of incentive packages grew more controversial.

By 1990, most counties and cities had accepted economic development as an important function of local government. Some vested responsibility for economic development in a public agency; others turned responsibility over to nonprofit development organizations. As separate local development commissions or as part of local Chambers of Commerce or neighborhood associations, these public agencies and nonprofit organizations had a mandate to promote local economic development in the face of dwindling resources and complex economic problems. In the 1990s, the escalation of social tensions, the reduction of economic security, and the continued physical and economic deterioration of many urban areas have returned distributional issues to the economic development arena. Yet, given the politics of fiscal austerity and the economics of increasing global competition, redistributive strategies have had little broad appeal. Promoting economic growth and development for the entire community is the preferred objective; increasing competitiveness remains the preferred strategy.[2]

Inadequate Explanations

Theories and models of local economic change that are intended to help practitioners understand the differences among places in the United States often are like products with very short life cycles. In the 1960s, urban economists argued that, because of powerful urban economies of scale and size, metropolitan areas would continue to enjoy competitive advantages and experience growth at the expense of nonmetropolitan areas. The "urban-size rachet" concept suggested that large places would continue to succeed. Over the business cycle, large places would hold their own better than smaller, more volatile places, resulting in a higher base from which to expand during the next recovery. By the end of the decade, the economic dominance of large urban areas and the relative poverty of rural areas were widely accepted ideas.

Analyzing the 1970 census results, U.S. Department of Agriculture demographer Calvin Beale (1979) documented a reversal of the trends. Nonmetro-area growth had surpassed metro-area growth, and a "rural renaissance" apparently had begun. Perhaps, bigger no longer meant better. As if to underscore this point, the essays of E. F. Schumacher (1973) in *Small Is Beautiful* became influential. Urban and regional theorists responded to Beale's evidence of relative urban decline with rationalizations about the diseconomies of scale and size. The relative decline of traditional industrial centers meant the diffusion of growth to other areas. The South and West grew much more rapidly than the North and East. Actually, this movement of population and economic activity represented nothing new; since the first U.S. decennial census in 1790, the center of the U.S. population has been moving southward and westward.

By the beginning of the 1980s, everyone was comfortable with the ideas of the rural renaissance and the dominance of the Sunbelt over the Snowbelt, often pejoratively called the Rustbelt. Again, the conventional wisdom was proved wrong, as rural manufacturing and resource-oriented areas went into recession, along with many parts of the Sunbelt. When shadows appeared in the Sunbelt and bright spots emerged elsewhere, researchers responded with the image of the bicoastal economy: prosperous regions bordering the two oceans and recession rampant in the nation's midsection.

By 1990, it was clear that this image again had been overtaken by reality. New England and numerous other parts of the country, including California and Florida, had come to experience the rolling recession that rumbled through the Southwest and farmbelt earlier in the 1980s. Declining defense industries were one proximate cause. Although the recovery and expansion of the 1980s did not take on a distinctive spatial pattern, California and other earlier "winners" continued to suffer more than other locations, and earlier laggards from the "oil patch" areas of Texas, Oklahoma, and Louisiana and the Rocky Mountain states grew at the highest relative rates. Thus, as a general rule, it appears that simplified ideas and explanations of urban and regional development prove largely incorrect by the time they take hold in people's minds. The spatial mosaic of growth and decline will undoubtedly continue to defy conventional explanations.[3]

Economic Development Practice in the 1990s

Development practice is exceptionally challenging and diverse. The issue of global competitiveness remains on the front burner while, in many places, the pendulum is swinging back toward broader social and environmental issues. Economic developers continue to focus on industrial recruitment and existing industry programs, but at the same time they embrace policies that promote entrepreneurship, small business development, tourism, and other potential sources of job growth. Enhancing the local economic base often involves direct assistance to promote exports. Looking within the locality for sources of growth is as common as looking to external sources.[4]

Although economic development practice has improved as it matures, it suffers from certain confusions and unanswered questions. First, economic development practice remains dominated by technique. Developers are involved primarily with marketing the locality, developing infrastructure, assembling sites, financing businesses, arranging skills training, and providing tax incentives. Economic developers remain enchanted with "doing deals," rather than pursuing a broader vision of economic development in the public interest (Jeep 1993). In essence, some developers fail to understand why and for whom they are working. Analysis and strategic thinking are still undervalued. This book offers economic developers theories and ideas they can use to guide analysis and inform strategic thinking.

Second, the definitions, goals, and strategies of economic development often are discussed without explicitly recognizing the relevant theory or model of economic development on which they are based. Without an explicit model of development, even the goal of job creation remains unclear. For example, should all permanent, full-time jobs be considered equivalent, or should differences in job quality be factored in? Should developers count all jobs or only net new jobs? Should jobs that reduce unemployment or accrue to existing local workers be valued more highly? Is the profile of jobs anticipated over time from a project important, or only the initial number of jobs? Should developers be concerned with local benefits exclusively, or should they be concerned with benefits that accrue to the larger economic system (Malizia 1987)? As will be shown in chapter 2, each theory provides both an explanation of the development process and a definition of what development is. Together, they suggest what needs to be done to promote development.

Third, economic development thinking often is based on weak economics. It still suffers from both mercantile and physiocratic biases. Modern economics notwithstanding, developers continue to confuse "making money," in terms of a favorable balance of trade generated by primary or manufacturing activities, with creating wealth by improving productivity. The concepts presented in this book should help resolve confusions and strengthen economic thinking in the field generally.

Fourth, development thinking is often ahistorical, which leads to gross blunders in formulating development strategies. Economic developers frequently ignore the economic history of the community in which they work. For example, in the late 1960s, the Appalachian Regional Commission, in devising its development strategy, neglected to examine the previous one hundred years of Appalachian history. The commission's strategy was based on the assumption that Appalachia was poor because it was isolated; therefore, it needed to be linked to the rest of the United States. The commission used growth centers, development highways, and social service programs to integrate the region into the larger economy. Yet there is evidence that the historical linkages between the region and the rest of the country have impoverished it. In a more recent example, Porter (1994) calls for the private sector to assist both new and existing inner-city businesses as a way to address central city decline. He does not, however, address the issue of how, over the past century, these areas became poor, dangerous, deteriorating places. The answers should affect what type of policy will be successful. The importance of history and human agency is underscored in chapter 2.

Finally, practitioners emphasize economic growth more than economic development, or they fail to recognize that growth and development are neither synonymous concepts nor identical processes. Jeep (1993) contends that the inability to distinguish growth from development has led developers astray. Growth and development are distinguished and contrasted in chapter 2.

To have a chance to succeed, economic developers need strong leadership, patience, persistence, and good planning. But all too often, they put their faith in good timing and luck to bring about local economic development. As William Alonso (1990) has said, somewhat in jest: we believe that a "silver bird" (manufacturing facility) will land on our island (local industrial park) and bring us development or that, through "spontaneous combustion" (local creativity and innovation), new local businesses will transform poverty into plenty. This book should help developers answer some basic questions and find ways to promote development by using their minds, rather than depending on magic.

Discussion Questions

The questions that follow are directed to economic development professionals. If you are a student or a reader in another field, you may use these questions to interview a practicing developer. The questions are designed to address important orientations of local economic development practice that bear on the definitions of development and the theories meaningful to developers.

1. Briefly describe the activities that take most of your work time over a calendar year. Who are the clients and major beneficiaries of these activities?
2. Identify the major strategies or policies that require these activities. Who are the primary beneficiaries of each strategy or policy?
3. Is job creation or investment for tax base expansion the primary goal of these strategies or policies? Do you have other primary or secondary economic development goals?
4. If job creation is an important goal, are you simply trying to stimulate the local demand for labor, or are you also trying to provide jobs for local residents or unemployed residents?
5. Economic developers have been criticized for engaging in constant-sum activity by increasing local growth but not overall national growth. Do you think this criticism applies to your activities? What is your view on this issue?
6. The outcomes listed below could occur as a result of your activities. Assume that each group generates the same number of jobs or amount of economic growth. Indicate those you would view as representing the best improvement to the local economy.
   (a) new branch facility / expansion of existing business / new locally based business
   (b) manufacturing company / retail company / business or professional service company
   (c) more small businesses / more self-employment
   (d) larger export base / cheaper products for local consumers
7. Do you address environmental quality or community development issues in your practice? How are these issues dealt with?

Please refer to the answers to these questions as you read this book.

Notes

1. Further historical background is provided in chapter 2.

2. Jeep (1993) and Rubin (1988) offer critical assessments of economic development practice, while Sternberg (1987) presents a detailed classification scheme for organizing economic development policy instruments. Smith and Fox (1991) describe

the major tools used in industrial recruitment, namely infrastructure provision, preferred financing of capital facilities, customized training, and tax breaks or financial incentives. They call for additional strategies to promote business development, including small business development centers, research and development partnerships, technology transfer through the manufacturing extension service, and new venture financing through venture funds.

3. One reason these popular explanations do not hold up to scrutiny is that they measure growth and development for the wrong economic units: states, multistate regions, or rural areas compared to urban areas. In chapter 2, we argue that the metropolitan economy or labor market area (the nodal region) is the best unit of analysis to use to measure economic growth or analyze the development process. Metropolitan economies in the continental United States, Canada, Alaska, and Hawaii, and, increasingly since NAFTA, in Mexico, are part of a North American market that is part of the larger international market. Furthermore, less developed countries with dollar-denominated currencies function more like regional economies within the North American market and less like foreign countries.

4. The American Economic Development Council (AEDC) describes current practice as pursuing "ACRE" in order to achieve more local jobs, investment, and tax base: attraction of new investment or facilities, creation of new businesses, retention of existing businesses, and expansion of existing businesses. Detailed descriptions of these activities can be found in Smith and Ferguson (1995).

2

Definitions and Concepts of Development

Ideally, a review of various theories of local economic development should be informed by a clear definition of the economic development process we seek to understand. Yet, an appropriate a priori definition of economic development is not possible. Definitions do not precede theory; each definition contains its own implicit theory, just as each theory supports a unique definition.

Definitions

Various definitions of "development" have had a guiding influence on development practice, usually without being explicitly connected to their theoretical basis. Three definitions influential in international development practice are presented to illustrate this point.

Before 1950, international economic development was identified with colonialism. The development process was imposed on undeveloped countries and expected to benefit the countries engaged in the exploitation. Initially, economic development involved the extraction and exploitation of undeveloped countries' natural resources for the benefit of the colonial powers. Subsequently, colonies served as production sites and markets for final products. Economic development, in this context, could be defined as the diffusion of development through production and exchange from more-developed to less-developed economies. The theories of the day agreed on most aspects of the diffusion process but differed sharply on its economic impacts.[1]

After 1950, with colonialism dismantled, economic development was supposed to benefit the less-developed countries. Public- and private-sector activities were designed to increase investment that would industrialize and modernize the economies of these countries. Economic development was defined as increasing per capita income and gross domestic product. Neoclassical growth and trade theories were consistent with this definition.

By the end of the 1960s, economic growth rates had been disappointing in some countries and impressive in others. It had become clear that, even in prospering economies, most people were not enjoying the benefits of growth. Therefore, Dudley Seers (1969) posed a new definition. He argued that economic development must create jobs, reduce absolute poverty, and increase income equality. His definition justified initiatives aimed at encouraging rural development, which became the dominant international strategy during the 1970s. Broader institutional theories, which focused on the structure of economies in less-developed countries, became associated with Seers's definition.

Development practice in the United States accepts economic growth as a positive force and attempts to facilitate the growth process.[2] In 1990, the American Economic Development Council, one of two major professional associations representing U.S. economic developers, commissioned a report from the profession entitled *Economic Development Tomorrow* (AEDC 1991). A Delphi process involving the profession's leadership, as well as other experts, accepted the definition presented in a similar report:

> *Economic Development:* The process of creating wealth through the mobilization of human, financial, capital, physical and natural resources to generate marketable goods and services. The economic developer's role is to influence the process for the benefit of the community through expanding job opportunities and the tax base. (AEDC 1984, p. 18)

This definition captures the two aspects of the term *economic development*. It refers to both a process and a practice. The economic development process is viewed as a growth process—the mobilization of resources to produce marketable products. The definition is static, however. It fails to indicate that economic development, as a process and a practice, is a long-term, ongoing enterprise. New development problems continue to emerge as former ones are resolved. Though flawed, the definition is nonetheless powerful because it justifies much of what practitioners now do in the name of economic development in the United States.

The definition has been referenced since the mid-1980s without addressing the fact that it is inconsistent. The first sentence is contradicted by the second. The first sentence is a politically astute description of the market system with which we are all familiar. Resources are mobilized and used to produce commodities for which market demand exists. Such a development process increases wealth because aggregate consumption increases. The second sentence instructs the economic developer to facilitate the development process through the creation of additional jobs and the expansion of the tax base. Yet, whereas wealth may be created and jobs/tax bases expanded simultaneously, they are fundamentally different objectives. Wealth creation results from the production and sale of commodities that benefit consumers. Job and tax base creation make available the human and physical resources needed to produce commodities; labor and government services are means to the end of greater consumption. The critical omission in the definition is the notion of scarcity, the existence of which dictates that jobs/tax base creation may erode, rather than generate, wealth. Wealth creation and jobs/tax-base expansion do not necessarily go hand in hand.

The existence of scarcity is a fundamental part of the economic process; scarce resources are used to satisfy competing ends. The economical use of resources is valued because it achieves efficient, least-cost production. Therefore, more consumption produced by less labor and fewer government services benefits the community, while more employment or tax revenues without more consumption impose unnecessary costs.

Although it is clear and easy to understand, AEDC's definition obscures important questions about contemporary economic development practice. For example, the definition does not address the serious criticism that developers frequently use scarce public resources to move jobs from one place to another, without contributing to national competitiveness (the constant-sum recruitment game). Moreover, the definition does not address whether developers should attempt to retain companies that want to close or relocate.

Furthermore, wealth creation benefits corporate shareholders and other business owners; job creation benefits local workers and community residents. While both are legitimate political objectives, they also are potentially conflicting. Trade-offs frequently arise between the development of successful companies and the achievement of attractive places to live. We can view these objectives as part of the debate between *equity* (stable and plentiful employment in many communities) and *efficiency* (national productivity growth). The AEDC's consensus definition is politically acceptable in part because it ignores these trade-offs. It optimistically implies that the engine of economic growth will create jobs and wealth where developers want them.

Economic developers have been more comfortable facilitating the economic growth process and working to improve the local business climate. They have been less comfortable trying to increase per capita income or wage rates. Nor has priority often been placed on equity issues or the problems of low-wealth residents.

A more logical definition of economic development, one consistent with mainstream (neoclassical) economics, would consider wealth creation more important than job creation—or, alternatively, job growth as a means to creating wealth. Local economic developers would only facilitate jobs and tax base expansion locally when wealth was not diminished or productivity reduced in the larger economic system. If taken seriously, this revised definition could significantly influence the practice of economic development.[3] Yet, more fundamentally, each theory provides the basis for posing a defensible definition of economic development.

Concepts

In order to learn how to use theory effectively in economic development practice, economic developers need to grasp four fundamental concepts. They face an apparent contradiction, given the conventional views of theory and practice. Theory is considered abstract thinking that simplifies some aspect of reality; practice involves human action that changes reality in some particular way. Social theory generally addresses "social forces" and ignores individual action to explain reality; it is essentially deterministic. Economic developers, as well as other individual actors, exercise free will in trying to influence the development process. Together, the concepts of power, theory, interests, and mediation resolve the apparent contradiction between deterministic theory and voluntaristic practice.

Power

Power is the ability to do work, to get something you want accomplished. Power is exercised in one of four ways: by using (1) money, (2) force, (3) persuasion, or (4) information. With money, people can buy what they want or hire other people to get what they want done. The government has a monopoly on the legal use of force, but some individuals use force illegally, if not illegitimately, to further their interests. Obviously, people do certain things because they are threatened with the use of force. Those who argue well can employ persuasion to get others to help them accomplish their objectives. Historically, a few individuals with great charisma have mobilized many people to work in their service.

These forms of power clearly have little to offer typical economic developers. Their access to money and the government's police power will depend on the organization and jurisdiction in which they work. Charisma varies with personality. Information, however—in the form of good ideas, facts, or knowledge—may be used to accomplish development objectives. Although it is probably the weakest form of power, practitioners can use information to better understand local realities.

Theory is the most powerful form of information. It can be used by developers to increase the power of information at their command. With theory, they will not only be better able to convince others to help them fulfill their plans, but they will also have the ability to think independently and creatively about the local economy in the face of political pressures favoring particular development strategies.

Theory

Theory usually is distinguished from action (practice). But the generic definition of theory implies action. Contemplation (a mental viewing) and speculation (formulating a plan) are mental activities that help prepare us to get something done. Theory is a systematic set of propositions positing cause–effect relationships and placed on a continuum from hypothesis to law. It can be contrasted with practice that is action-oriented, not reflective. Yet the more basic idea is that theory offers the underlying principles that explain the relationships we observe and thereby motivates and informs our action.

The concepts of theory and power can be brought together by picturing people doing things, thereby exercising their power. They use theory to guide their actions in order to produce and reproduce reality. Theory as an abstract, static system of causes and effects is data—that which is given. Theory used by people in practice results in facts—that which is made. The Latin roots of the words *data* and *fact* make this distinction. Again, facts are made, usually the result of human action. In other words, facts are mediated by human agency and informed by theory.

Like other professionals and businesspeople concerned with practical affairs, developers often scoff at theory as useless abstraction. Yet, to paraphrase John Maynard Keynes, such practical minds usually are preoccupied with the ideas espoused by some dead and largely discredited economist. Everyone operates with a theory or model of reality, whether they recognize it or not (Boulding 1956).[4]

Interests

Many local and nonlocal actors influence the local development process. Their influence is unequal because they have access to different forms and degrees of power. In a democratic, market-oriented society, developers have neither the authority to dictate development strategies nor the capital needed to expand production directly; instead, they must rely on persuasion to carry out strategies. In other words, development practitioners must listen to (and understand) powerful actors in order to communicate with them effectively and motivate them to act in ways consistent with petitioners' programs. Therefore, developers must understand the theories and models used by other participants in the development process. Attaining this understanding is a primary justification for studying theory.

Of course, theories and models are not politically neutral. Rather, people tend to believe that which is in their best interest and ignore conflicting evidence. In other words, they are opportunistically informed and opportunistically ignorant. Will Rogers is said to have observed that the problem of ignorance is not caused by people who don't know anything. The problem is caused by people who know things that ain't so.

A convenient way to identify economic interests in development is to recognize the four general impacts the development process has on different actors: price, quantity, income, and/or wealth effects. Theories that refer to the economic growth process focus on income effects and quantity effects (such as changes in employment or output). Local actors often feel price and wealth effects more immediately and more acutely. For example, consider the large discount retailer who comes to an area, bringing employment and earnings opportunities (quantity and income effects). As a result, several local businesses are forced to lower their prices (price effect) and become marginal enterprises. Landowners near the new retailer realize capital gains on their property; retail property owners in other parts of the community experience capital losses (wealth effect). All these effects deserve attention if the costs and benefits of growth are to be understood in a comprehensive manner. In considering the applications of each theory, the developer should explicitly consider the relationship of the theory to local political and economic interests by outlining the likely price, quantity, income, and wealth effects.[5]

Mediation

Human agency is people planning and doing the things which represent the specific activities and events that affect the course of development in specific places. All cause-effect relationships considered in social science are mediated by human agency. Yet social scientific theories are largely deterministic explanations of human behavior; otherwise, they would not be simple, logical explanations of reality. Theory-based analysis treats observed facts as data. Theory that considers voluntaristic human activity is not typically part of social science; rather, historical studies embrace human action by describing who did what, when, and where, under which circumstances and to what effects. For example, historical studies document the tremendous impact Robert Moses had on the growth and development of New York City and its environs. For the sake of simplification and generality, however, development theory applied to New York City effectively ignores Moses' role, focusing instead on an explanation of the movement and interaction of broad economic indicators.

Developers should pay attention to theory to understand reality and avoid the trap identified by Keynes. Like other men and women of action, developers believe in their power to change future reality. They believe that economic development does not just happen, that people make development happen. *Develop*, then, is a transitive, not an intransitive verb (Arndt 1981). To overcome the apparent contradiction between trying to understand the world deterministically and exercising free will while trying to change it, informed practice must place in historical context the human activity that led to the observed facts.

In general, economic development theories ignore human agency. Theory lays out cause–effect relationships without addressing the activities of people required to give life to these relationships. In abstract causal models, inanimate objects appear imbued with human traits and abilities. Economists usually speak of "the market" as if it had a life of its own, as if it could *do* things. Meanwhile, they ignore the human agency required to make markets function.

Unfortunately, human action usually is assumed away in development theory. The practitioner wishing to use theory must always keep in mind the notion that all theoretical propositions are mediated by human agency. Without some degree of "reading between the lines," the full picture will remain obscure. One must realize that, in the interest of conciseness, human agency is often ignored.[6]

As noted, historical studies underscore voluntaristic actions when they describe who did what, when, and where. Historians are the researchers who document human agency with all its messiness in their scholarly work. Historical analysis, though not theory, provides the indispensable background and context that lead to wise, informed application of theory. Unfortunately, social scientists rarely relate their behavioral theories to the human action described by historians. Thus, without both theoretical explanation and historical analysis, our understanding of economic development is incomplete. Local economic developers who understand the role of mediation should be able to illuminate the relationship between theory and practice and place them in historical context.

Although we focus on theory and its application to practice, the other three concepts should be remembered when considering each theory presented in Part II. *Power* raises questions about how practical the theory is and how easily and usefully it can be applied. We expect developers to be drawn to theories that are both insightful and appropriate given the local context. *Interests* motivate the developer to gauge, from the perspective of each theory, the benefits and costs of development and to identify the winners and losers. *Mediation* requires careful study of a location's economic and political history as necessary preparation for any serious attempt to use theory to influence the development process.

Basic Assumptions

The assumptions described in this section provide the conceptual basis for the economic development theories presented in this book. We assume that the local economy is best understood within the context of the larger economic system. The evolution of the local economy can be grasped by understanding its relative attractiveness compared to other local economies. This attractiveness depends on economic location—the role the location plays in the larger system.

One viable path to economic development is to create and build links between the local economy and the external economic system. Another calls for strengthening internal links and minimizing external trade—what is called *autarky*. If the local economy is to achieve economic progress following the former path, it must find a useful role to play in the global economy. Over time, successful local economies will specialize, trade, grow, diversify, and develop. If the local economy is to progress following the latter path, it must develop through local self-reliance.[7] For many developing regions and countries, both paths deserve consideration, but for most U.S. localities, building links to the global economy is more feasible and desirable (if not unavoidable) than separation and autarky. The audience for this book, along with its regional focus, suggests an emphasis on *competitiveness* as the primary objective by which to evaluate the relative success of linkage to the global economy and the relative attractiveness of the locality to firms and households. Localities are drawn into interaction by centripetal economic forces, which are usually more powerful than centrifugal political forces.

The global economy is viewed as a system that consists of mutually exclusive metropolitan areas or, more broadly, local labor market areas functionally linked through economic exchange—that is, through the flow of goods, services, money, credit, information, and people. Metropolitan areas are the most appropriate unit of analysis with which to study relative attractiveness because they represent meaningful economic entities and functional economic areas. Each area has one or more functional specializations. In this book, the terms *locality, area, region, location, place,* and *community* generally refer to local labor market areas.

Because the United States is a fairly mature economy, our coverage of regional development theory selectively includes macro-oriented theories that emphasize the diffusion of development. The focus of Part II is on the continued diffusion of development across local economies, rather than the initiation or early stages of development.[8] Micro-oriented theories of spatial development are not covered. Migration theory, which deals with population dynamics and movements that result from the decisions of households, is largely excluded. Location theory, as it pertains to decisions concerning business location, also is generally ignored because, in this book, space is treated discontinuously as discrete places rather than continuously as the "friction of distance." However, to the extent that location theory offers useful insights about the competitiveness of places, it is incorporated in the argument (see chapter 8).

Growth-Development Distinctions

The understanding of local economic development can be sharpened considerably by examining the distinctions between economic growth and economic development. Robert Flammang (1979) identifies nine different implicit or explicit conceptions of economic growth and development adopted by researchers and practitioners. They may be summarized as follows:

1. No definitions are offered. Economic growth and development are reduced to other concepts such as urbanization and industrialization.
2. Growth is the same as development. Growth, or development, is measured as an increase in aggregate or per capita income.
3. The distinctions depend on geography. Growth occurs in rich countries, development in poor ones.
4. The distinctions depend on the origin of development. Change that comes from sources internal to the region is development, change that is externally imposed is growth; yet sometimes the argument is reversed.
5. Growth and development are complements because one makes the other possible.
6. Growth and development are alternating processes that occur in sequential time periods.
7. Growth is an increase in output; development is structural change—technical, behavioral, attitudinal, or legal.
8. Growth expands the economy; development must lead to more equal distributions of income and wealth.

9. Growth or development leads to a greater range of economic choices.

Flammang presents an ecological model to synthesize and describe long-term change. To better adapt to their environments, populations are organized into societies. Development is triggered when a population begins to crowd its environment. Out of necessity, the society draws on its environment in an attempt to increase its means of sustenance. Adaptation involves seeking a new ecological niche. From the ecological perspective, development involves niche-finding; growth is niche-filling. Organized populations (communities) introduce adaptive technology to solve economic problems.[9] This adaptation may or may not lead to adoption by the ecological system. If adaptation occurs, the population is able to increase; otherwise, population stagnation, decline, or outmigration occur.

Using this context, Flammang argues that growth is best defined as simple, quantitative increase, whereas development is more qualitative and involves structural change.[10] Growth and development may be competitors in the near term but are usually complements in the long term. Over the long term, growth provides the resources needed for development; development generates new technical, organizational, behavioral, or legal structures that facilitate growth. Growth increases output by mobilizing more resources and utilizing them more productively; development changes the output mix by devoting local resources to doing different kinds of work. In the near term, however, growth or development may proceed without the other; growth may retard development or development may engender decline. Moreover, development can occur in one place by draining resources from another location, thereby limiting growth and development elsewhere. Regional disparities in growth rates and development levels are common features of the growth-development process.[11]

The growth-development distinction also suggests counterposing economic growth to *sustainable* development. Indeed, sustainability has become a widely discussed principle in the economic development field. The concept is attractive because of its generality; it would appear to encompass issues of competitiveness, social inequality, and environmental quality. Most of the literature, however, presents normative statements outlining sustainability principles or descriptions of specific experiences, particularly in developing countries. Although cogent theories of sustainable development have yet to be formulated, the basic idea is to evaluate economic growth in view of its impact on people and nature. The criterion appears to be to economize in physical and resource terms. Thus, economic growth with the following features would be preferred:

1. Limited utilization of natural resources and, in general, respect for ecosystems;
2. Efficient utilization of material resources through energy conservation, recycling, and so on;
3. Full use of existing capital stock and reuse of sites and buildings;
4. Less pollution from production and consumption activities; and
5. Reasonable levels of living with less income and wealth inequality.

Although the concept remains too broad to be usefully applied in this book, some suggestive literature is worth noting.[12] In particular, the work of Karl Polanyi deserves brief attention. Polanyi (1944) developed a unique perspective on economic development that was drawn from careful historical and anthropological analysis. He argues that traditional (precapitalist) societies subordinate the economy to their politics and culture. Capitalist development represents the historical anomaly. Under capitalism, the economy becomes the dominant societal force. Ultimately, when human beings and the natural world are treated as mere commodities, the competitive market, which is a highly effective mechanism for allocating produced commodities, becomes a destructive mechanism. If people and nature are reduced to the commodities of labor and land, the economy is likely to damage the society it was expected to serve.[13]

From the outset, people have tried to avoid the free market, seeking instead monopolistic harbors to protect their sources of income and wealth. Businessmen continually try to circumvent the market through price fixing and other forms of commercial collusion. Workers form unions "in restraint of trade." Environmentalists fight the reduction of natural resources to material production inputs. Government intervention attempts to balance the needs of the market economy to treat people and nature as commodities with the political reaction against such treatment.

Polanyi would ask economic developers to find the balance. Developers can be decidedly "pro-business," advocating measures to improve the local business climate on the grounds that private investment is needed to sustain the local economy. Nevertheless, developers should recognize the need for intervention that reasonably protects workers and the natural environment from market forces. Although this is much more difficult than one-sided advocacy, developers should address both the costs of growth and its benefits.

Theories

To apply a theory successfully, the economic developer must understand its language. This section summarizes each theory in terms of five fundamental elements (see table 2.1):

1. *Basic categories*—the fundamental classification or distinctions used to lay out the theory
2. *Definition of development*—what economic development is or should be according to the theory
3. *Essential dynamic*—the key variable or relationship that drives the logic of the theory
4. *Strengths and weaknesses*—how well the theory enables one to understand economic development
5. *Applications*—the ways in which the theory can be used in economic development practice

The basic categories of *economic base theory* are the industrial sectors of the regional economy assigned to either the basic sector or the nonbasic sector. The definition of local economic development is equivalent to the rate of local economic growth measured in terms of changes in the local levels of output, income, or employment. The essential dynamic of the theory is the response of the basic sector to external demand for local exports, which, in turn, stimulates local growth. The economic base multiplier transmits change in output, income, and employment from the basic sector to the entire regional economy. The theory's major strengths are (1) its popularity as a basis for understanding economic development in North America, and (2) its simplicity as a theory or tool for prediction. Its major weakness is its inadequacy as a theory for understanding economic development, especially in the long term. Economic base theory strongly supports attracting industry through recruitment and place marketing.

Staple Theory

Staple theory identifies industrial sectors as its basic categories. It defines economic development as sustained growth over the long term. The essential dynamic is the external investment in, and demand for, the export staple that leads to the successful production and marketing of the export staple in world markets. The theory's major strengths are its historical relevance to North American economic development and its emphasis on understanding the region's economic history. Its major weakness is that it describes, more than explains, the development process. Staple theory provides a general strategy of development by recognizing the connections of the economic base to the political superstructure. Economic developers should continue to build on and improve the export staple as long as it remains competitive in the larger economic system. The idea is to "stick to one's knitting," since strengthening the existing specialization may be more sensible than attempting to diversify the economic base. Eventually, footloose economic activities (that is, those not closely tied to *specific* resources, inputs, or markets) will be attracted to the area if its market achieves sufficient size or if it offers urbanization economies that can be exploited by other exporters.

Sector Theory

Sector theory uses three aggregate sectors as basic categories. The level of development depends on sectoral diversity, emphasizing a prominent tertiary sector, and labor productivity. The essential dynamic involves the income elasticity of demand and labor productivity of primary and secondary sectors: as incomes rise, the demand for income-elastic products grows; output increases as labor released from primary and secondary sectors is employed in tertiary sectors. Although sector theory is attractive because it can be applied and tested empirically, the primary, secondary, and tertiary categories are too crude to be useful in practice. The overriding application is the need to attend to industries producing income-elastic commodities in order to achieve sustained growth.

Growth Pole Theory

Growth pole theory treats industries as the basic unit of analysis, one that exists in an abstract economic space. Economic development is the structural change caused by the growth of new propulsive industries. Propulsive industries are the poles of growth, which represent the essential dynamic of the theory. Growth poles first initiate, then diffuse development. Growth pole theory attempts to be a general theory of the initiation and diffusion of development based on François Perroux's domination effect. Although insights drawn from the theory are useful, it has failed as a general theory of development. Growth center strategies are based on this theory. Also summarized in table 2.1 are the growth theories of Gunnar Myrdal and Albert Hirschman, which are consonant with Perroux's theory.

Neoclassical Growth Theory

The basic categories of neoclassical growth theory are sectors or regions that comprise the macro economy. Economic development is defined as an increase in the rate of economic growth, measured in terms of changes in output or income per capita. The theory has two essential dynamics. One, in aggregate models, the rate of saving that supports investment and capital formation drives the growth process. Two, in regional models, factor prices—specifically, the relative returns on investment and relative wage rates—stimulate factor flows that result in regional growth. Growth theory suggests that economic developers respect the free market and do what is necessary to support the efficient allocation of resources and the operation of the price mechanism. The simplest growth models imply that economic developers are unnecessary, but more complex formulations would support various economic development activities.

Interregional Trade Theory

The basic categories of interregional trade theory are prices and quantities of commodities and factors of production, just as in microeconomics. The implicit definition of development is economic growth that leads to greater consumer welfare. The essential dynamic is the price mechanism (price-quantity effects) operating to eliminate price differentials and establish equilibrium prices (the terms of trade). The theory has two unique strengths. First, consumer welfare (increases in aggregate consumption benefits), not job creation, is the goal of development. Second, the price/cost-based theory is extremely precise, yet its precision is achieved with numerous restrictive assumptions and largely by ignoring the dynamics of development. Economists use growth theory and trade theory to advocate less government intervention and freer international trade, more open regions, and, in general, more competitive markets. The theories provide strong support for local infrastructure development, improvement in government efficiency, and other measures that could increase local productivity and lower input costs for all producers. Local developers, on the other hand, often ignore the implications of growth and trade theory and instead support protectionist measures and growth strategies that do not always improve the economic well-being of local consumers.

TABLE 2.1

Summary of Economic Development Theories

Theory	*Basic Categories*	*Definition of Development*	*Essential Dynamic*	*Strengths and Weaknesses*	*Applications*
Economic Base Theory	Export or basic and nonbasic, local or residentiary sectors.	Increasing rate of growth in output, income or employment.	Response to external changes in demand; economic base multiplier effects.	Most popular understanding of economic development in the United States and a simple tool for short-term prediction. Inadequate theory for understanding long-term development.	Industrial recruitment and promotion for export expansion and diversification, expansion of existing basic industries, import substitution by strengthening connections between basic and nonbasic industries, and infrastructure development for export expansion.
Staple Theory	Exporting industries.	Export-led economic growth.	Successful production and marketing of the export staple in world markets. External investment in and the demand for the export staple.	Historical perspective on economic development. Descriptive theory difficult to apply.	Build on export specializations. State does everything possible to increase competitive advantage. Character of economic base shapes political and cultural superstructure.

Theory	*Basic Categories*	*Definition of Development*	*Essential Dynamic*	*Strengths and Weaknesses*	*Applications*
Sector Theory	Primary, secondary, and tertiary sectors.	Greater sectoral diversity and higher productivity per worker.	Income elasticity of demand and labor productivity in primary and secondary sectors.	Empirical analysis possible. Categories are too general.	Promote sectoral shifts. Attract and retain producers of income elastic products.
Growth Pole Theory	Industries.	Propulsive industry growth leads to structural change.	Propulsive industries are the poles of growth.	General theory of initiation and diffusion of development based on the domination effect.	Growth center strategies.
Regional Concentration and Diffusion Theories	Commodities and factors (Myrdal) or industries (Hirschman).	Higher income per capita.	Spread and backwash effects (Myrdal) or trickle-down and polarization effects (Hirschman).	Address the dynamics of development.	Active government to mitigate backwash effects and reduce inequalities (Myrdal). Location of public investments spurs development (Hirschman).
Neoclassical Growth Theory	Aggregate (macro) or two-sector regional economy.	Increasing rate of economic growth per capita.	Rate of saving that supports investment and capital formation.	Supply-side model.	Government should promote free trade and economic integration and tolerate social inequality and spatial dualism.

TABLE 2.1 (continued)

Summary of Economic Development Theories

Theory	*Basic Categories*	*Definition of Development*	*Essential Dynamic*	*Strengths and Weaknesses*	*Applications*
Interregional Trade Theory	Prices and quantities of commodities and factors.	Economic growth that leads to greater consumer welfare.	Price adjustments that result in equilibrium terms of trade; price-quantity-effects.	Unique emphasis on consumer welfare and price effects. Ignores the dynamics of development.	Government intervention should promote free trade. Infrastructure development, efficient local government.
Product Cycle Theory	Products: new, maturing, or standardized.	Continual creation and diffusion of new products.	New product development; innovation.	Popular basis for understanding development among researchers.	Development strategies promote product innovation and subsequent diffusion.
Entrepreneurship Theories	Entrepreneurs or the entrepreneurial function.	Resilience and diversity.	Innovation process; new combinations.	Mediated theory.	Support industrial milieu or ecology for development.
Flexible Specialization Theories	Production regimes, industrial organization.	Sustained growth through agile production, innovation and specialization.	Changes in demand requiring flexibility among producers.	Detailed analysis of firm/industry organization; aggregate outcomes and relationships seldom specified.	Encourage flexibility through adoption of advanced technologies, networks among small firms, and industry cluster strategies.

Product-Cycle Theory

Product-cycle theory treats the developmental age of the product as its basic category. Products are classified as new, mature, or standardized. At any point in time, the space economy can be divided into regions where new products tend to arise and regions devoted to the production of standardized commodities. The essential dynamic of product-cycle theory is new product development, which is one form of innovation. From locations where new product innovation takes place, the product is eventually standardized and diffused to other locations in the space economy. The process stimulates economic growth and development in both types of locations, but the character of development is different in each. These differences help explain why levels of development vary from place to place, and why differences can persist. The economic developer who wants to apply product-cycle theory in its most literal form must try to identify and work with manufacturing companies that can create new products. Alternatively, the developer may be able to mobilize the resources needed to improve the local business infrastructure in ways that would support new product development.

Entrepreneurship Theories

The basic category of economic development is the entrepreneurial function as embodied in the entrepreneur. Development proceeds as changes in firms and industries result in more resilient, diverse local economies. The essential dynamic driving the development process is innovation. Innovation is conceptualized variously in different theories as new combinations, improvisation, or creative risk taking. To its credit, entrepreneurship theory is mediated theory; people make development happen. This strength, however, leads to the weakness that entrepreneurship theory is not easy to apply consistently. The most general application is to support an industrial environment or ecology favorable to entrepreneurs.

Flexible Production Theories

Flexible production theories focus on production regimes and related methods of industrial organization as basic categories. The regional development implications of customized, batch, and long-run (or "Fordist") production regimes—as well as outsourcing practices, supplier relations, and processes of vertical integration and disintegration—are the principal concerns. Development is not just quantitative growth but also qualitative change in industrial mix, firm structure, and sources of competitiveness (for example, from least-cost or price-focused competition to that based on innovation, product differentiation, and niche marketing). More recent research has focused on the impact of flexible production on labor practices, compensation, and power relations between large and small firms. The key variable or relationship (essential dynamic) that drives flexible production theories are changes in the nature of demand that require firms to become more agile; standardized, least-cost production is considered less and less viable as consumer tastes in industrialized countries become more sophisticated and global competition intensifies. Firms adapt to this new environment by adopting flexible production technologies, managing supplier relationships, and utilizing interfirm networks for information sharing and joint problem-solving.

Among the principal strengths of the theory are a focus on rich, complex production dynamics within firms, between firms, and between firms and labor. Weaknesses are related to the strengths in that the focus on specific micro relations means that implications for regional aggregates are often neglected. In terms of application, the theory informs industry cluster strategies, buyer–supplier networking initiatives, technology transfer programs, small-firm programs, and some types of worker ownership and labor management policies applied at the community level.

Summary

Each theory of development described in this book can be used to examine economic development in two distinct ways. First, a theory can be used to arrive at the practitioner's own understanding of the realities of the economic development process. Second, it can be used to understand how others may be thinking about development. In an attempt to influence the development process, practitioners must appreciate the consciousness and interests of all relevant actors. Embedded in every theory is a particular definition of economic development, implicit goals, and development strategies, as well as its implications for the wealth and income of various actors.[14]

Not surprisingly, no single theory explains the economic development process adequately. Many theories require study: classical, neoclassical, Keynesian, institutional, Austrian, and Marxian. In fact, these theories represent different paradigms or schools of thought. Each paradigm uses a different language; both the rules of grammar and the vocabulary are different. Because of this, we attempt to present each body of theory in its own terms rather than defining and criticizing concepts from a single point of view. This approach is consistent with the view that no one theory or paradigm can be expected to help the practitioner understand every development situation or solve every development problem. A variety of theories together offers the best way to understand both the reality of, and the thinking about, development.

This book, then, is written to help economic developers with two vital tasks: first, to understand the reality of local economic development more clearly, in order to achieve greater material well-being for their communities; and second, to understand the thinking of other actors about economic development, both to win powerful supporters of the developer's program and to communicate strategies to the general public. In the first instance, the developer gains insights about *what* to do; in the second, he or she begins to figure out *how* and *with whom* to do it.

Discussion Questions

1. What is power? Which forms of it are most useful to economic developers?
2. How can the concept of theory be related closely to and motivate human action?
3. Do you agree that everyone operates with a theory of reality?
4. Why is it important to understand interests in economic development?
5. How can theory help developers understand interests?
6. Why is it difficult to mediate social science theories?
7. Why is it necessary to supplement economic development theory with historical analysis?
8. How would you define economic development? What does your definition imply for both the process and practice of economic development?
9. Compare and contrast your definition of "economic development" to the AEDC's definition.
10. How does your definition of "economic development" address these trade-offs:
    (a) developing successful companies versus achieving an attractive place to live;
    (b) wealth creation for corporate shareholders/business owners versus the benefits to local workers and community residents;
    (c) stable and plentiful local employment versus increasing national productivity growth.

The following questions refer to Appendix 2.1.

11. Are the ideas of mercantilists and physiocrats alive and well in the

minds of many local leaders, or do most agree with the classical thinkers led by Adam Smith and therefore think more like contemporary economists?

12. Which set of ideas is the most influential? The following questions should help identify important elements in order to contrast preclassical to classical economic thought.
    (a) Is wealth created by accumulating money, by increasing value added in one sector, or by increasing productivity throughout the economy?
    (b) Is manufacturing (or agriculture or mining) the wealth-creating activity, whereas other economic activities are wealth circulators; or are all forms of economic activity valuable?
    (c) Should government assist producers to increase the balance of trade, or should consumer interests be more important?
    (d) Should government generally try to minimize its role in the economy and focus on efficient delivery of services, or should government intervene to stimulate private investment?

Appendix 2.1

ECONOMIC THOUGHT

Economic thought is presented in its historical context, for four reasons. First, it is the written product of scholars arguing with each other about the realities of economic life. The arguments began in earnest in the eighteenth century and continue today. This give-and-take cannot be grasped by studying economics in standard textbooks. Second, because thought and practice continually interact, by understanding historical economic thought, the reader can correlate concepts and arguments to the major historical events of the period. Ideas are formulated to explain changes in economic life, and, to some extent, economic life is shaped by the accepted economic doctrines of the time. Third, the historical sequence of the theories presented in Part II helps one understand them more fully (Heilbroner [1972, 1988] and Galbraith [1987]). Fourth, and most importantly, the economic thought covered below can help economic developers understand much current thinking about the development process.

The outline of economic thought shown in figure 2.1 presents the general flow of ideas over the past 250 years. From the many early ideas, mercantile and physiocratic thought are highlighted because certain long-standing arguments about economic development in the United States are based on

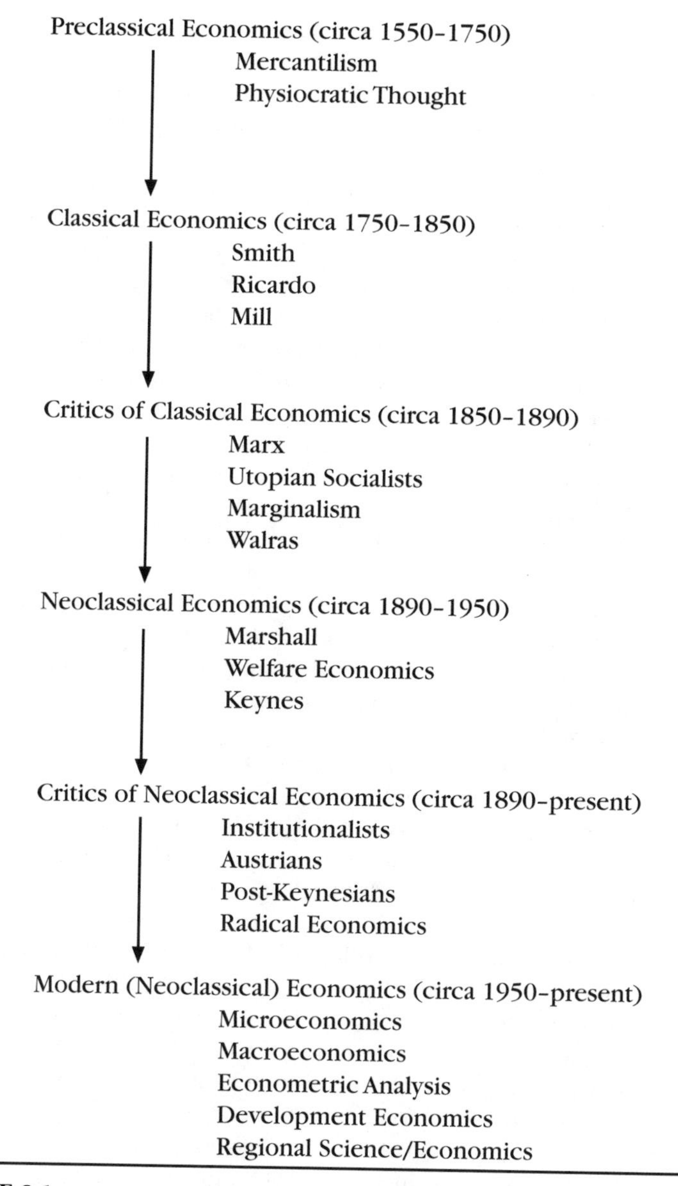

FIGURE 2.1
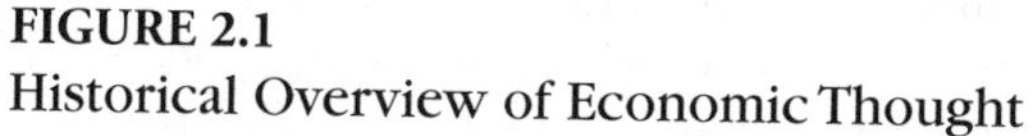
Historical Overview of Economic Thought

these ideas. Thus, mercantile and physiocratic thought, which most contemporary economists hardly know and never use, continue to influence contemporary economic development practice. Adam Smith, David Ricardo, and John Stuart Mills, writing in that order and building on earlier work, present the first complete treatment of classical economics (political economy). The critics of classicism include both supporters of capitalism—the marginalists and Walras—and dissenters in Marx and the socialists. Alfred Marshall consolidates marginalism and classical thought and combines them with the theory of supply and demand. Over the next several decades, many others, including Keynes, broaden and deepen economic theory. These ideas are synthesized in modern micro and macro economics, which, with the support of mathematics and econometrics analysis, continue to grow. Neoclassical critics have generally objected to the formalism and level of abstraction in modern economics. Critiques of the institutionalists and Austrians have waned, while critiques of post-Keynesian and Radical economics continue.

As Arndt (1981) points out, development economics did not become a recognized subfield of economics until the end of World War II. It tries to grasp the evolution of the economy in terms of economic growth and structural change and draws theories from the body of economic thought portrayed in figure 2.1. Classical thought is particularly relevant. Unlike neoclassical theory's concern for equilibrium or steady-state growth, the major objective of classical thought is to understand the evolution of the economic system over the long term.[15]

The spatial dimension of structural change is largely neglected in the economic thought portrayed in figure 2.1. Economists from Great Britain were concerned, first, about national material progress achieved over time and, later, with economic statics and dynamics. Economists on the European Continent took up issues of space, regions, and urbanization. The theory of location was almost entirely developed by German and Central European economists from about 1820 to 1940. Isard (1956) synthesized this work, which subsequently spawned the subfields of regional science and regional economics.

The theories presented in Part II are rooted in different streams of thought, since regional economic development cannot be addressed adequately by one theory or school. Economic base theory (chapter 3) is compatible with both regional and Keynesian theory. Staple theory (chapter 4) is a regional application within the institutional tradition. Perroux, Myrdal, and Hirschman (chapter 5), early critics of mainstream development economics, had their ideas incorporated into early formulations of regional economics. Neoclassical growth and trade theories (chapters 6 and 7) flow most directly from mainstream neoclassical economics as applied in development economics. Post-Keynesian theory (chapter 6) provides insights about growth and trade that contradict neoclassical outcomes. Product-cycle theory (chapter 8) is an interesting combination of neoclassical and regional theory. Schumpeter's theory of entrepreneurship and innovation (chapter 9) belongs to the Austrian school. Radical and Marxist theory inspire the literature on flexible production presented in chapter 10. While theories in Part II provide insights about the *process* of local economic development, the remainder of this appendix is devoted to explanations and applications of economic thought that should help clarify the way many people *think* about local economic development. The thought is presented in historical sequence.[16]

Mercantile Thought

Mercantilism is not a theory of economic development but a diverse set of ideas about the economic development process and government's role in that process. Mercantile thought was prominent in Western Europe for several hundred years through the middle of the eighteenth century. Its ideas were based on poorly defined terms and were not consistently applied. Perhaps the continuing appeal of elements of mercantile thought stems from the fact that mercantilists were men of business and practical affairs. Though insightful, they certainly were not theoretically inclined.

Mercantile thought made the pursuit of wealth respectable. Progress based on exchange was a radical departure from the medieval view of commerce. The church saw "usury" and many other business practices as evil. Mercantilism helped elevate secular power at the expense of religious influence. State power was seen as the force that fuels economic growth. Mercantile thought was most prominent in England, which, during the seventeenth and eighteenth centuries, became the dominant European power.

Mercantilism was centrally concerned with trade: nation-states should achieve and retain a favorable balance of trade.[17] With merchandise exports greater than merchandise imports, money in the form of gold or other precious metals was received from foreigners to pay their trade-related debts. Precious metals were accumulated in the national treasury and used to fund state spending. Thus, rulers of the nation-state and the merchant class were allies. The merchant class helped provide the funds needed by the nation-state to increase its power for consolidating the national territory, colonizing peripheral areas, or waging war. The international trading companies received state protection and favorable government policies, and the merchants got rich in the process.

Mercantilists argued that, through effective state intervention, growth could continue in the foreign trade and manufacturing sectors, resulting in the accumulation of public and private wealth. They ignored domestic demand and the difficulties of structural change, instead focusing on the exchange of goods for money. Mercantilists proposed (and debated) far-reaching interventions to support foreign trade: navigation laws, tariffs on imports, direct export promotion, state-protected trading monopolies, product monopolies (patents), transportation and other infrastructure investment, and so forth.

Mercantilists believed in the strategic importance of commerce and industry, preferably controlled by relatively large companies. Aggregate production could be increased by expanding the labor force through immigration and natural increase and by using better capital and more skilled labor. Sufficient money supply would keep interest rates low and encourage domestic industrial growth by stimulating the demand for labor and other inputs. The growth process would increase the surplus available for foreign trade when (1) resources were used productively, (2) agriculture provided cheap wage goods, and (3) wages remained at subsistence levels. National power would increase with the growth of output and population, as well as with the accumulation of money wealth.

Applications of Mercantilism

Mercantile ideas continue to have broad appeal because they appear sensible and connected to economic reality. Indeed, people who believe that politics dominates economics are especially attracted to mercantile thinking. Regional economists have extensively studied the influence of federal and state policies on regional development. Markusen (1986) has traced the regional influences of defense spending, while others have looked at the political economy of U.S. trade (Noponen, Graham, and Markusen 1993). Certainly, economic developers need to be aware of the local impacts of state and federal policies and expenditures.

State and local public officials and economic developers who focus on subnational economies cannot use many of the interventions proposed by the mercantilists because these are available almost exclusively at the national level. Their ideology or strategic orientation is often consistent with mercantilism. Many developers support the process of economic growth that leads to greater aggregate output, jobs, and tax base. Developers are less comfortable facilitating increases in per capita income, higher wages, or better working conditions.

Like the mercantilists, many developers believe in interventions that are pro-business, that are intended to support and facilitate business plans. Importantly, they do not value all firms equally. Large oligopolistic companies in the foreign trade (basic) sector are the most important. These companies generate the income that supports all local exchange. However, economic history teaches that oligopolies frequently are undermined and that competition rules in many sectors of the economy where higher levels of consumer well-being often result.

The mercantile idea that a favorable balance of trade is essential for economic prosperity has misled economic developers more than any other idea. This misconception leads many developers to concentrate on the process of making money rather than increasing productivity. A favorable balance of trade can help sustain local growth, but internally focused exchange can do so as well; yet many developers are keen on increasing the locality's trade surplus. Most believe in export promotion, primarily by attracting export industries to the locality, and in import substitution that reduces income leakages. Attraction of exporters and import substitution probably are the most popular development strategies in the United States.

This mercantile misconception can be corrected, as follows. Regions and nations need to sell products in international markets in order to buy products in these markets without incurring ever-increasing debt. But accumulated funds from trade, in themselves, indicate neither wealth nor prosperity. Rather, prosperity and wealth come from the productive deployment of local resources. A favorable balance of trade can indicate economic strength, but this simply means that the value of exports exceeds the value of imports. This situation will continue as long as local companies can compete successfully in external markets. The presence of strong (efficient or flexible) companies is what causes economic growth; net exports are simply an indicator of growth.

Physiocratic Thought

The physiocrats represent the first true school of economic thought. Writing in the eighteenth century, they proposed a reasonably coherent, consistent body of ideas about the political economy of France. Decidedly antimercantilist and ardently laissez-faire in spirit, the physiocrats tried to show objectively how different parts of the economy interrelate and how value is created and circulated. Although many of their tenets were rejected by Adam Smith and the classical economists, they set a high standard for economic thinking that continued to challenge classical economists.

The French economy, largely agricultural, was suffering, and the landowners and peasants were burdened by heavy taxes and harmful government policies. The physiocrats favored an enlightened French monarchy. They wanted to preserve and reform France's agricultural system, save the landed aristocracy that owned it, and reduce the excesses of the French court. They opposed imposition of mercantile policies on the French economy and the growing influence of the merchants and the manufacturing sector.

Physiocratic appeals for economic reform proved insufficient; the French Revolution swept away the privileged classes and restructured the French economy.

The physiocrats believed that all wealth came from nature, embodied in land and natural resources. Agriculture and other resource-based activities, therefore, created wealth; all other forms of economic activity merely circulated wealth. The productive classes included landowners and agricultural managers. Merchants, professionals, artisans, manufacturers, and bureaucrats were considered unproductive classes. The labor class was dependent on these other classes for employment and was either productive or unproductive, depending on the type of employment.

Physiocratic thought was inspired by rationalism and natural law. Property rights and free trade were supported. Except for national defense, any intervention on the part of the state was suspect. Government intervention to further mercantile interests was ardently opposed. Government revenues, needed for a limited number of activities, were to be raised by direct, single taxes on the agricultural surplus that was the income of the productive classes. By simplifying the tax system, the size of government could be more easily reduced.

The physiocrats devised an ingenious accounting framework to underscore their position on productivity. This framework is often viewed as the forerunner to input-output economics. The "tableau economique" presented by François Quesnay attempted to measure the "net product" or surplus. The table was used to show how the net product circulated through the economy. After compensating the productive classes who worked the land, the net product belonged to the aristocracy.

The physiocratic conception of value directly opposed the mercantile definition of wealth. Trade added no value because the commodity traded was merely circulated, not changed. Nor did manufacturing add value since industrial labor simply worked on materials originally from the land. Economic progress could be achieved only by increasing agricultural production and thus the agricultural surplus. Agriculture, therefore, was the strategic sector, not just the sector that limited manufacturing growth.[18] Income per capita could grow by allocating sufficient capital to agriculture, by allowing competition and free trade, especially of agricultural products, and by serving the growing domestic market. In contrast to mercantilism, exports and balance of payments, colonies, and expansionary population policies were not important.

Applications of Physiocratic Thought

One enduring, useful idea from the physiocrats is the interconnectedness of the economic system. Not only are different local economic activities interdependent, local economies themselves are interconnected. As noted, input-output analysis is a means of tracking important links through commodity flows in the local economy. The concept of linkage is an important one for economic developers. Analysis of it helps them understand how the local economy connects to the larger system and how local sectors support that connection.

Although economic developers do not accept the physiocratic conception of value, many believe in the distinction between productive and unproductive economic activity. Basic economic activities—namely, manufacturing, extraction (mining, forestry, and fisheries), and agriculture—are wealth creators. Other forms of economic activity merely circulate wealth. This idea is wrong; *any* economic activity that satisfies consumer wants and needs creates value.

More generally, economic developers should favor no sector or set of firms on a priori grounds—for example, basic activity rather than local activity, manufacturing rather than services, large firms rather than small firms (or the reverse). None is inherently superior. In healthy local economies, for example, growth of manufacturing and services tends to be complementary and reinforcing. Instead, developers should try to identify strategically important sectors and firms. These sectors have high multipliers, provide good jobs, offer developmental services, export income-elastic products, and consist of efficient or flexible firms. The theories presented in Part II define strategic importance in meaningful terms.

Physiocratic thought is strongly identified with the economic interests of local property owners, just as mercantilists favor merchants and industrialists. Large property owners often support less government regulation of business and defend economic freedom and property rights. Many believe that taxes on business income and wealth should be minimal. Whether consciously or not, economic developers generally support the economic interests of groups that directly benefit from aggregate economic growth. Representatives of these groups dominate the boards of economic development organizations and are allies of economic developers.

Classical Economics

The famous Scottish moral philosopher Adam Smith presented the first complete treatise on what became known as classical economic thought (political economy) in *The Wealth of Nations* (1937 [1776]). Yet this book was actually based on an ethical system developed in Smith's earlier work, *The Theory of Moral Sentiments* (1976 [1759]). As a moral philosopher concerned with human happiness and well-being, Smith presented the moral sentiments that would restrain narcissistic human tendencies. The sentiments of sympathy and the need for social approval, he argued, would balance self-love and provide the moral basis for economic affairs. With this moral and ethical framework in place, Smith's "simple system of natural liberty" would harmonize self-interest with the public interest and keep competition humane.

Self-interest motivates and directs economic activity which, in turn, is regulated by competition. Smith defends the competitive capitalistic system as an efficient way to achieve secular progress. The "invisible hand" of the market leads to "natural" prices and just compensation of labor, land, and capital. The cooperative efforts of labor employed on farms or in factories produce useful goods. It is natural for people to specialize in what they can do well and trade for what they need. The division of labor increases productivity. Specialization, ultimately limited by the size of the market, increases the size of the pie, and competition insures that consumers can purchase at a reasonable price the commodities for which they are willing to pay. The nation prospers under this system of natural liberty.

Smith's factors of production also represent social classes—land (property owners), labor (workers), and capital (business owners). He was particularly concerned about the well-being of workers who, in the transition from feudalism to capitalism, had lost the economic security of being tied to the land.

Smith poses new definitions of value and wealth. Value is created neither from accumulated money nor from agricultural surplus, but from productivity—that is, the ability to combine labor, land, capital, and other inputs efficiently so as to produce more output. Wealth is increased by the continual division of labor, which leads to specialization and trade. Free trade not only ensures the best allocative outcomes, it also increases the size of the market, which, in turn, stimulates further specialization and division of labor.

Although known for his support of laissez-faire, competition, and free trade, Smith articulated important roles for government, much larger than those the physiocrats would have accepted. Besides enforcing the legal framework for contracts and private property, the government should regulate, tax, and spend in the public interest. Regulation is justified when the social costs of private activity exceed the private costs. Public production is necessary when the social benefits exceed private benefits such that no rational private producer would provide the good or service.

Smith devoted considerable attention to criticizing mercantilism. He considered mercantile thought to be a fraud perpetrated on the public by businessmen. He demonstrated why the accumulation of money was not the wealth of the nation and railed against the monopolistic trading practices of the mercantilists.

Smith was more favorably disposed to the physiocrats. He shared their views about productive and unproductive classes and the existence of natural order. Yet there were sharp and important differences. According to Smith, natural liberty resided in the people, not some enlightened despot. Value came from human effort more than from the gifts of nature.

Smith presented a creative, readable work that influenced economists over the next century. He distinguished his basic tenets from mercantile and physiocratic thought and considered competitive capitalism to represent the natural order of the long-term development process. Smith's concern for economic development in England remained the major focus throughout the classical period.

The most important classical thinkers after Smith were Ricardo, Malthus, Say, and J. S. Mill, all of whom tried to develop a set of laws describing the natural order of capitalist development. Ricardo focused on income distribution and foreign trade. Ricardo formulated two important contributions—comparative advantage and diminishing returns. His concept of comparative advantage, developed to explain specialization and commodity trade between countries, is fully explained in chapter 7, which covers interregional trade theory. He was the first to see that specialization could lead to diminishing returns. In his pessimistic scenario, in contrast to Smith's optimistic growth model, the economy eventually stagnates, and landowners become the dominant class. Expansion through trade and colonialism are ways to continue growth and postpone eventual stagnation. This scenario underestimates the ways in which technological progress and substitution can overcome diminishing returns. Figure 2.2, adapted from Dome (1994), neatly portrays the contrasting classical views of economic growth formulated by Smith (expansion of market, increase in profits) and Ricardo (decrease in profits).

Malthus is best known for his work on the influence of population increase. He presented the most cogent, classical statement on the limits of economic growth. In essence, the rate of national economic growth depends on the relationship between population growth and the growth of capital stock. The argument continues today with debates about economic growth, population increase, natural resource constraints, and environmental deterioration.

Say transformed Smith's ideas to textbook format for French language readers and made several contributions in the process. He is best known for Say's Law, which argues that markets should always clear because the compensation of factors producing supply provides the demand for that output.

J. S. Mill wrote the leading English language treatise on economics in the latter part of the nineteenth century. He synthesized and contributed to classical thought and incorporated social justice issues raised by socialist critics

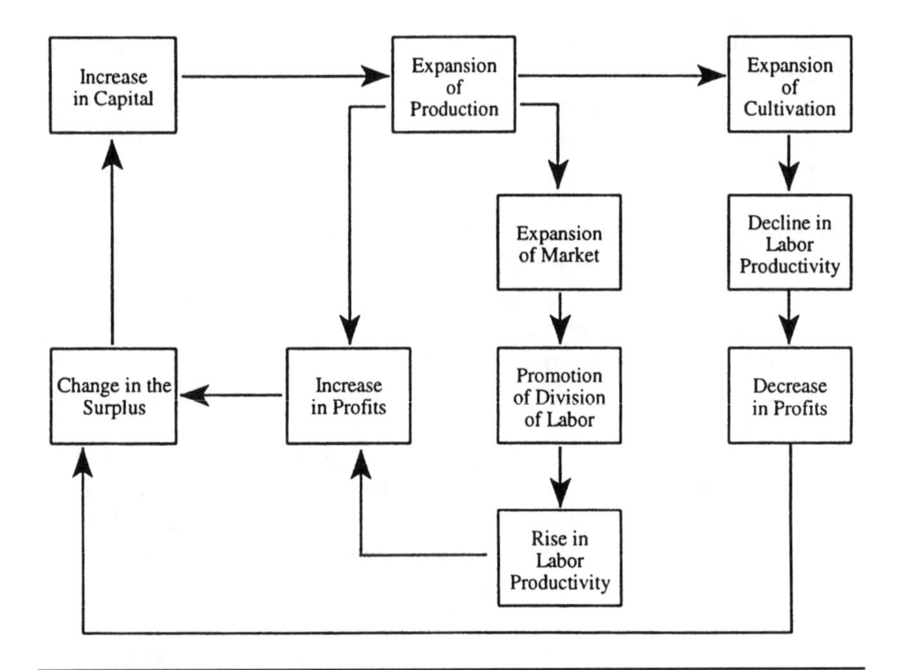

FIGURE 2.2
Classical Economic Growth Process

Source: Figure adapted from Dome (1994).

of capitalist development. Mill became an advocate of "population" policy (encouraging emigration), land for the poor, public education, and even women's rights. He thought these measures would reduce population growth and elevate the living standards of workers above the subsistence level (Dorfman 1991).[19]

Applications of Classical Thought

Classical economists developed economic concepts that contemporary developers can use. They are particularly useful in countering preclassical economic thought. As noted, many people appear to accept mercantile ideas and, to some extent, physiocratic ideas, and they behave as if these ideas were valid. They are influenced by economic thinkers long dead and largely discredited, probably because these ideas serve their narrow economic interests.

Although modern economics has gone beyond Smith and other classical economists, certain classical tenets remain relevant to clear thinking about economic development. Smith would want developers to focus on competitive firms, and he encouraged government to support competition. He would strongly advocate greater specialization to serve viable local and external markets. He would want places to build on their specializations to sustain productivity growth. He would advocate trade to satisfy local demand, and more exporting and importing, rather than less.

Classical economists would not be especially supportive of economic developers and politicians who simply wanted more local economic activity and greater creation of local jobs. They were primarily interested in near-term efficiency, long-term productivity gains, consumer satisfaction, and the wealth-creation process. Economic developers who embrace these ideas would promote broader community interests. Others who remain comfortable with preclassical ideas would promote narrower business and property interests.[20]

Marx

Classical economists were concerned primarily with understanding the evolution and dynamics of capitalist development. In this sense, Karl Marx was a classical economist. He drew directly on Ricardo, both in substance (value theory) and in method. Yet, Marx was a broad social thinker more than an economist. He attempted to synthesize and build on Hegel's historical method, physiocratic thought, and socialist ideas.

Marx presented his version of "scientific socialism" to demonstrate how and why capitalist development will fail. He argued that the collapse of capitalism, and the subsequent rise of socialism, are caused by internal contradictions. That capitalist development is inherently unstable, rather than adhering to Say's Law, is one of Marx's most important contributions to economic thought. Another is his emphasis on the importance of technological progress. His insightful theory of capitalist political economy, however, is inconsistent and incomplete.

The most pertinent area of Marxist thought is the definition of social classes. Marx used economic function to determine class position in keeping with the classical approach (land, labor, and capital refer to three distinct social groups). The upper class owns the means of production; the working class is employed by others and owns little beyond housing and personal property. The managerial class runs private and public organizations in the interest of the upper class. The proprietor class represents small business. The managerial and proprietor classes own some productive assets, but much less than the upper class. Finally, the unemployed own very little and may even have negative net worth (Bottomore 1966).

Applications of Marxist Thought

Economic developers working in the United States will find few, if any, local actors who think about development from a Marxist perspective. Instead, the main application of Marxist thought is in linking what people *say* about development to what they *do* (to their class position). Marxist thinking can help developers be aware of competing and conflicting economic interests in the locality, as well as the groups that are influential in the political arena because of their control of capital. Developers should recognize that the strategies they support favor certain economic interests more than others, often those of business owners and managers.

Economic developers are expected to be savvy about the politics of their community in order to facilitate positive changes in the economy. Leadership roles in the largest organizations headquartered in the region and the individuals occupying these roles should be identified and their affiliations determined. The channels and extent to which these people exercise political leadership should be understood. The objective here is not to discover some "ruling class" that uses democratic structures to impose its singular will on the populace; rather, analysis should help developers determine the strength of business influence in local governance and the relative difficulty involved in implementing an economic development strategy.

Often, members of established families not only play key roles in local production but shape the local culture by filling other leadership positions. Together, these families run the large local companies; the banks that finance these companies; the law firms and other business service firms that have these companies as clients; the newspapers, radio stations, and other local media; and possibly even influential civic, religious, and educational organizations.

The presence of a coherent, influential class varies from region to region. In some communities, the political system is quite open and the process pluralistic and democratic. Elsewhere, political discourse and processes are tightly controlled. Economic developers need to learn quickly the kind of environment in which they are working. Neither extreme ensures a favorable outcome, however. In relatively democratic places, it is fairly easy to raise issues, discuss alternatives, and mobilize enthusiasm for various development strategies; yet it may be impossible to reach agreement on an overall strategy or to mobilize the resources needed to implement the strategy. In less pluralistic places, power is sufficiently concentrated for taking effective action and mobilizing resources. But it may be next to impossible to convince the leadership of the need for change, or it may take too long to get their attention.[21]

Economic developers enjoy little security of tenure. They must be able to analyze the politics of the place rapidly and accurately. They should seek to include people who know the local politics in their professional network. Access to this knowledge has saved developers lots of time and has prevented them from stepping on political land mines.

In summary, early economic thought can be very useful to economic developers trying to understand local thinking about economic development. Mercantile and physiocratic thought are at the root of popular misperceptions about economic development. Classical thought can help correct these misperceptions while describing economic development in ways more likely to serve the public interest. Marxist thought should help developers see the connection between local views of economic development and the economic interests served by these views.

Notes

1. The classical economists identified mutual benefits associated with bilateral trade. The Marxists developed the theory of imperialism to explain the process. Further discussion of historical economic thought is found in Appendix 2.1.

2. More formally, information on alternative locations for production facilities is imperfect. Developers help perfect the market by improving the quality of information used in investment decision-making. This information is provided by marketing the locality and its sites. Economic development practice as a form of place marketing is well described in Kotler (1993).

3. A more detailed version of this section is found in Malizia (1994), whose revised definition of economic development states:

> [T]he ongoing process of creating wealth in which producers deploy scarce human, financial, capital, physical and natural resources to produce goods and services that consumers want and are willing to pay for. The economic developer's role is to participate in the process of national wealth creation for the benefit of local consumers and producers by facilitating either the expansion of job opportunities and tax base or the efficient redeployment of local resources. (p. 84)

This critique and revised definition apply neoclassical economics with which most readers are familiar. As noted, other definitions of development can be drawn from other theories and paradigms. The definitions from theories covered in subsequent chapters are summarized in table 2.1.

4. The sequence of discussion questions in chapter 1 presents a way to grasp the essential features of a developer's theory. First, the developer is asked to describe what he or she does. Next, the activities are understood as means to accomplish one or more development strategies. Third, strategies are examined to identify the developer's definition of economic development. As noted, each definition of development implicitly assumes some theory of development.

5. Joan Robinson and John Eatwell relate theory to interests in an interesting way, which previews the discussion in Appendix 2.1:

> As we have seen, the Mercantilists were the champions of the overseas trader; the Physiocrats supported the landlords' interest; Adam Smith and Ricardo put their faith in the capitalist who makes profits in order to reinvest them and expand production. Marx turned their argument round to defend the workers. Now, Marshall came forward as the champion of the rentier—the owner of wealth who lends to the businessman and draws his income from interest on loans. (1973, p. 39)

By extension, Keynes appreciated the consumer and government as major forces influencing the business cycle, whereas Schumpeter emphasized the power of the entrepreneur as producer. All theories have their champions, as we demonstrate in Part II.

6. It may be useful at this point to practice the application of this admonishment by analyzing the unmediated statement—"costs increased, thereby causing prices to rise"—and writing down the actions of workers, unions, suppliers, business executives, and influential groups (such as OPEC) that would be required to realize this simple causal statement. After sufficient practice, one will better understand that all causal relationships postulated by a particular theory are mediated. One can also see why the specific activities of economics actors are too cumbersome to be fully described in behavioral models.

7. Agropolitan development, parallel economy, and basic-needs approaches have been proposed to build self-reliance and are discussed more fully in Friedmann and Weaver (1979). This path requires that political and legal institutions either shield the local economy from external forces or regulate economic forces in order to protect people and the natural environment.

8. An excellent comparative treatment of the origins of capitalist development in various nation-states is in Moore (1966).

9. Flammang's ideas about development have much in common with those of Thorstein Veblen. Both view technological progress as a common property resource that is used to increase the society's survival potential. This idea may be counterposed to Schumpeter's individualistic conception of the entrepreneur who leads the development process (chapter 9). Veblen advanced a critique of capitalist development rooted in cultural and historical analysis of modern societies. He viewed economic progress as driven by "instincts" conditioned by culture: a desire to be productive (workmanship), responsibility for friends and family, and concern for the next generation (parental bent), which, respectively, lead to regard for quality, community, and the future. With the rise of corporate enterprise, however, business (or commercial/finance) principles come to dominate industry (production) principles. Corporate financiers, absentee owners, and top managers gain power at the expense of engineers, technicians, and lower managers. Salesmanship becomes more important than workmanship. As a result, productive capacity is not fully used, and prices and profits are increased at the expense of greater production at lower cost.

10. The analogy of human growth and development is appropriate. We all grow, and eventually shrink, as we age. We also develop as we mature from infant to toddler, to youngster, to teenager, to adult, and to elder. These terms refer to qualitatively different developmental stages.

11. In a later article, Flammang (1990) skillfully contrasts growth to development with the idea of static efficiency (growth) versus dynamic efficiency (development). He argues that development "softens" economic and social structures, which leads to greater flexibility, whereas growth "hardens" structures in order to realize efficiencies. For any local economy, growth involves decreasing internal differences, increasing structure, accumulating rewards, and expanding to fill existing opportunities. Development involves increasing internal differences, flexibility, and adaptive capacity while searching to find new opportunities that respond to external change. These distinctions can be applied by thinking about local companies that exemplify the following differences. A growing company concentrates resources on several promising product lines or services, organizes production to increase margins, and gains a greater share of existing markets. A developing company establishes new profit centers, decentralizes control, experiments with new business approaches, and pursues new markets or different marketing channels.

12. There is a long tradition of posing strong critiques of capitalist development in response to its deleterious environmental and social impacts. The writings of Peter Kropotkin, Patrick Geddes, and Lewis Mumford come to mind. In addition, more recently, a reformulation of economic theory in biological terms has occurred (for example, see Daly [1991]).

13. The two questions that concerned Polanyi are: (1) how did the free market become so dominant, and (2) how have people and governments responded to its dominance? The shorthand answer to both questions is an epigraph to an article about Polanyi: laissez-faire was planned; planning was not (Sternberg 1993). The self-regulating market remained dominant for about a century, rising to prominence from 1832, with the repeal of England's Elizabethan poor laws and passage of laws supporting the market, and remaining until 1870, then declining until the repeal of the gold standard and the trade wars of the 1930s. In the early nineteenth century, national governments created the institutional framework to support laissez-faire, which included the legal system to defend property rights and enforce contracts, the production of money and regulation of the money supply, the provision of necessary public goods, and so forth.

Polanyi reviews, from the early nineteenth century, the various and consecutive social reform movements that arose spontaneously to protest laissez-faire and call for the protection of labor and land from the self-regulating market. His interpretation of capitalist development and the need for social intervention represents a middle ground, one that is less extreme than that of Marxists who want to replace the market system, yet much more interventionist than the defenders of laissez-faire would tolerate.

14. The connections between theory and the interests of various actors in the development process are discussed in Malizia (1985, pp. 35-36). The reader should make these connections with *every* theory presented in Part II. Each theory would impose different benefits and costs on local interests. Price, quantity, income, and wealth effects are likely to impact local interest groups differently.

15. Meier (1984) distinguishes three types of thought used in development economics: analytical, radical, and historical. Analytical thought is most prominent and

involves applications of neoclassical growth and trade theories. Radical theories, such as dependency theory, have been formulated with the realities of less-developed countries in mind and are used to view development from the less-developed–country perspective. Historical theories, which include the work of Marx and Schumpeter, focus on long-term growth and change. In an earlier work, Meier's development theories are classical, Marxian, neoclassical, Schumperterian, and post-Keynesian; see Meier and Baldwin (1957). Herrick and Kindleberger (1983) devote two chapters to theories. In chapter 2, they review growth theories from the classical, neoclassical, and post-Keynesian traditions; in chapter 3, they present theories of economic development: neoclassical (Bauer, H. J. Johnson), structural disequilibrium (Chenery), and Radical (Amin, Baran, Gunder Frank). See also Ranis and Schultz (1988) and Stern (1989).

16. The main sources used in this review are Rima (1991) and Spiegel (1983). Additional references include Bronfenbrenner (1979), Chong-Yah (1991), Dome (1994), Dorfman (1991), Heilbroner (1972, 1988), Friedmann and Weaver (1979), Galbraith (1987), Herrick and Kindleberger (1983), Landreth and Colandeer (1989), and Russell (1945).

17. The balance of payments framework described in Appendix 3.3 should help the reader follow this discussion of international trade.

18. Strategies to increase economic development often favor industrialization—the growth of manufacturing. Manufacturing growth was limited by available inputs, the most important of which was labor, since the large majority of the population was engaged in agricultural production—often at the subsistence level. An increase in agricultural productivity was needed to generate surplus wage-goods and surplus labor. This reasoning is consistent with sector theory, presented in chapter 4.

19. Rima presents a concise list of the laws developed by the classical economists, as follows: "(1) the law of diminishing returns, (2) the law of population growth, (3) the law of wages, (4) the law of capital accumulation, (5) the law of rent, (6) the law of comparative advantage, (7) the law of value, (8) the quantity theory of money, and (9) the law of markets" (1991, p. 189). Note that these laws do not account for technological change that was emphasized by Marx (Dorfman 1991).

20. The contemporary moral basis of capitalism deserves some attention. In our increasingly mobile, technologically sophisticated society, it is difficult to find meaningful community life. Yet meaningful communities would appear to be a prerequisite for providing the social experiences that lead to the sentiments of sympathy and approbation. Although we may behave with regard for others who live in our neighborhoods, the most positive sentiment we seem capable of mustering is tolerance. With the loss of Smith's moral sentiments, economic activities and political affairs are likely to become less ethical and, all too frequently, illegal.

21. Often, Tupelo, Mississippi, is cited as a place where political leaders have mobilized for effective action in the interest of local economic development. It would be worthwhile to explain the proximate causes of success. See Holladay (1992), Martin (1994), and the *Wall Street Journal* (March 3, 1994).

PART II

Theories of Economic Development

3

Economic Base Theory

Economic base (or export base) theory dominates the thinking about local economic development in the United States. Its underlying premise—the external demand for a region's products as the primary determinant of regional prosperity—is widely accepted. Economic developers should treat learning economic base theory as a top priority, for a knowledge of economic base theory will enable them to understand and communicate better with other economic development practitioners and political leaders.

As a basis for understanding the reality of local economic development, however, economic base theory offers limited useful insights. In the form of a quantitative model, it can be applied for impact analysis and for making predictions of economic growth only as long as the structure of the local economy does not appreciably change. For long-term analysis of economic development, economic base theory has limited application because key export sectors and local economic structure change over time, often significantly.

The economic activities of a labor market or metropolitan area are divided into those that produce for the export market (called "basic industries") and those that produce for the local market ("nonbasic," "service," or "residentiary" industries). The two sectors are linked in two ways. First, the basic sector directly purchases goods and services from the nonbasic sector. Second, workers employed in the basic sector purchase food, clothing, shelter, public services, and other commodities from the nonbasic sector. Greater demand for the region's exports generates export sales and income for the basic sector, while basic-sector purchases provide income to the nonbasic sector. Consumption spending by basic-sector workers also generates income for the nonbasic sector; nonbasic firms and workers then engage in additional rounds of spending. These rounds of spending generate what is called the "multiplier effect." Through the multiplier effect, an initial increase in basic-industry income generates even greater total income for the area. In parallel fashion, a decrease in basic-industry income leads to a greater decrease in total area income.

Industrial recruitment is the most popular economic development strategy in the United States. This strategy finds its rationale squarely in economic base theory because the practice of attracting promising manufacturing export industries to a region is justified with the assumption that growth in nonbasic or local-serving industries inevitably will follow. In broader terms, the economic developer's most important task is to make sure the region possesses a suitable proportion of industries whose products are in heavy demand from outside the region.

I. Economic Base Theory

Economic base theory originally was developed as a practical technique for evaluating the local economy. Subsequently, it was used by urban planners to predict metropolitan population growth.[1] It also constitutes the simplest formulation of what may be described as a Keynesian theory of regional income. The theory postulates a simple demand-driven model of regional (local or metropolitan) growth. The standard economic base model divides the regional economy into two sectors. The *export sector* includes all economic activities that produce commodities sold to nonlocal households, businesses, or governments; the *nonbasic sector* includes all economic activities that produce for the local market. In relation to Standard Industrial Classification codes (SICs), primary sectors (agriculture, forestry, and fisheries) and secondary sectors (mining and manufacturing) often are considered basic, whereas construction and tertiary sectors (distribution, trade, and services) are nonbasic. Bifurcation of the regional economy into the two sectors can be refined by using SICs at more disaggregated levels and basing the classification on both logic and empirical analysis. Appendix 3.1 summarizes various ways to bifurcate the regional economy and estimate the economic-base multiplier, which is the primary area of continuing research on the economic base model.

The simplest economic base multiplier model—measured in terms of employment, income, or output—can be written as follows:

$$b + r = n \tag{1}$$

where b, r, and n denote basic, nonbasic (residentiary or local), and total regional economic activity, respectively. Nonbasic economic activity is a function of total activity,

$$r = f(n) = an \tag{2}$$

where a is a parameter that must be estimated. The multiplier is derived by substituting an for r in (1) and rearranging terms to get the following equation:[2]

$$\left(\frac{1}{1-a}\right)b=n \tag{3}$$

The term "premultiplying b" is the economic or export base multiplier; a denotes the average (and, in this case, also the marginal) propensity to spend locally. Given an exogenous change in basic activity (employment, income, or output), caused by an increase in demand for regional exports, the model can be used to estimate the associated change in total regional employment, income, or output.

The economic base model provides a means of estimating the employment, income, or output multiplier effects of external changes in demand for the region's exports. As external demand changes, output, income, and employment in the basic sector fluctuate in response. The linkage between the basic and nonbasic sectors is represented by the multiplier $(1\text{-a})^{-1}$, which transmits changes from the basic sector to the entire regional economy. If we assume that the multiplier remains constant over time, then we can project employment, output, or income from estimated changes in the export sector alone. For example, in an urban area with a calculated multiplier of 5.0, an increase in basic employment of 100 workers would lead to a total urban-area employment increase of 500.

Economic base theory focuses exclusively on external demand as the determinant of regional growth. Internal demand and supply are viewed as relatively insignificant sources of income and employment growth. Other features of the regional economy—including its endowment of natural resources, the rate of capital investment, and the quality and quantity of the labor force—are not addressed. The theory also assumes no significant production capacity constraints, implying that there are unemployed resources (capital and labor) within the region or that these resources may easily be acquired (via migration) from outside.[3] Only if that is the case will growth take place without changing local factor prices. And, as is characteristic of numerous economic growth models, the theory implies that increases in output, income, or employment increase consumer well-being; increases in welfare are equated with growth.

Many of these assumptions are tenuous at best. As a result, they have been subjected to significant criticism, some of which is summarized in section III of this chapter. Yet, in many quarters, economic base theory retains its appeal. One reason is that the theory is strongly intuitive; it is difficult to imagine that output and income in a region could grow if regional industries merely "took in one another's laundry." Another reason is that the relevance of the theory for policy—assuming that it is an accurate model of the growth and development process—is apparent. The postulate—that metropolitan growth in income, employment, and output are determined by external demand for the region's exports—has clear implications for economic development practice.[4]

II. Applications

Although economic base theory, including its input-output extension, focuses considerable attention on exporting industries, external demand for basic sector goods cannot be influenced directly from the local level. As an alternative, developers can try to expand the existing economic base in three ways. First, they can engage in industrial recruitment and promotion, to diversify the export base. To do this, developers might attempt to identify national or global growth industries and assess the region's competitive position vis-à-vis other areas. Attractive industries are those that offer forward or backward linkages and thus generate high multipliers. Second, developers can directly facilitate the expansion of export industries already part of the basic sector. Third, they can indirectly facilitate export expansion by improving the efficiency of local public infrastructure and services.

Development practitioners can use economic base theory in more subtle ways.[5] First, developers may not want to assume, as does the theory, that there are no limits on further growth and expansion (capacity constraints). Instead, they could work to ensure adequate supplies of trained labor, industrial land and facilities, public infrastructure, affordable housing, and so forth, since capacity constraints of any kind could retard growth. Second, although industrial recruitment is a much more straightforward approach, developers could try to diversify exports by devising ways to stimulate the founding of new exporting enterprises. Third, developers could recognize that, due to interindustry linkages, all industries give rise to multiplier effects; therefore, the most rapid rate of growth is not necessarily achieved by concentrating development efforts solely on the basic sector. Careful study of the basic sector and its linkage within the regional economy could open up paths for stimulating growth. Strengthened linkages between industries in the basic sector and the nonbasic sector would increase multiplier effects and reduce import leakages. This strategy is essentially one type of import substitution.[6]

Aside from economic base theory, which offers ideas for designing development strategies, the formal economic base model is the foundation of several important economic analysis techniques, including impact analysis and population projection. In the next two subsections, we demonstrate how local area, economic base multipliers are developed, using metropolitan Columbus, Ohio, as an example. We also describe how developers in Columbus use an economic base logic in their industrial targeting effort.

Calibration of Economic Base Multipliers

Using the location quotient method of bifurcating the regional economy described in Appendix 3.1, we calculated economic base employment multipliers for the Columbus metropolitan area. We used employment data because employment is the most common metric for computing location quotients. The third and fourth columns of table 3.1 list 1989 total employment for forty-five sectors in the Columbus Metropolitan Statistical Area (Delaware, Fairfield, Franklin, Madison, and Pickaway counties) and in the United States as a whole; the fifth column lists each sector's location quotient. Location quotients below 1.00 suggest that the Columbus economy is less specialized in the given sector than the United States as a whole; for the purposes of bifurcating the economy, all employment for these sectors is considered nonbasic. Applying equation (5) in Appendix 3.1 to those sectors where the location quotient exceeds 1.00 yields the amount of employment in the given sector that is hypothesized to produce exports. The Columbus economy is more specialized in these sectors than in the United States; thus, under the location quotient method, we assume that these sectors produce more than the Columbus economy can consume.

The results of these calculations are provided in the third and fourth columns of table 3.2. Column totals indicate that 61,289 workers produce for customers outside the Columbus metropolitan area from the total 1989 workforce of 652,653. Applying equation (3) above where $n = 652{,}653$ and $b = 61{,}289$, the multiplier equals 10.65, which means that an increase of 1,000 jobs in the basic sector should lead to a total of 10,650 jobs in Columbus, including the original 1,000 basic-sector jobs. This rather high multiplier is not atypical for large, diverse metropolitan areas like Columbus that have significant employment in trade, finance, insurance, real estate, state and local government, and other business and personal services.

The location quotient approach provides a simple, commonly used way of gauging the impact of industrial recruitment efforts that target the basic sector. As applied, however, the method is extremely mechanical, and it pro-

TABLE 3.1

Columbus and U.S. Employment,
Location Quotients by Industry, 1989

		1989 Employment		
SIC	*Industry*	*Columbus MSA*	*United States*	*Location Quotient*
	AGRICULTURE	3,677	489,293	1.22
	MINING	978	710,593	0.22
	CONTRACT CONSTRUCTION	28,040	5,035,938	0.90
20	Food and kindred products	8,266	1,445,994	0.93
21	Tobacco products	0	45,610	0.00
22	Textile mill products	235	676,161	0.06
23	Apparel and other textile products	515	1,069,137	0.08
24	Lumber and wood products	1,111	713,546	0.25
25	Furniture and fixtures	1,949	521,516	0.61
26	Paper and allied products	1,829	631,614	0.47
27	Printing and publishing	8,789	1,543,632	0.93
28	Chemicals and allied products	4,483	858,271	0.85
29	Petroleum and coal products	150	113,587	0.21
30	Rubber and miscellaneous plastic products	6,391	903,855	1.15
31	Leather and leather products	375	124,854	0.49
32	Stone, clay, and glass products	7,175	527,295	2.21
33	Primary metal industries	2,645	746,556	0.58
34	Fabricated metal products	8,171	1,524,991	0.87
35	Machinery, except electrical	8,907	1,975,565	0.73
36	Electric and electronic equipment	7,304	1,613,673	0.74
37	Transportation equipment	3,215	1,854,515	0.28
38	Instruments and related products	4,719	1,000,627	0.77
39	Miscellaneous manufacturing industries	1,224	396,039	0.50
—	Administration and auxiliary	5,562	1,205,407	0.75
	TRANSPORTATION AND UTILITIES	32,530	5,417,515	0.98

vides a multiplier based on very little insight and absolutely no intuition about the particular metropolitan area. Economic developers should, at a minimum, use their knowledge of their local economy to examine for reasonableness each location quotient and subsequent estimate of basic employment. To make adjustments, they can examine employment and location quotients at different levels of detail (two-digit SIC, three-digit SIC, or four-digit SIC). In this case, by identifying and adjusting location quotients that probably underestimated the portion of basic employment in the sector, we generated a revised multiplier that yielded more realistic estimates of the jobs that would result

TABLE 3.1 (continued)

Columbus and U.S. Employment,
Location Quotients by Industry, 1989

SIC	*Industry*	**1989 Employment** *Columbus MSA*	*United States*	*Location Quotient*
	WHOLESALE TRADE	39,766	6,153,130	1.05
	RETAIL TRADE	128,679	19,335,163	1.08
	FIRE	55,303	6,801,616	1.32
70	Hotels and other lodging places	6,650	1,467,566	0.74
72	Personal services	7,251	1,142,074	1.03
73	Business services	39,309	4,749,349	1.34
75	Auto repair, services, and garages	4,669	842,792	0.90
76	Miscellaneous repair services	2,434	372,580	1.06
78	Motion pictures	935	347,623	0.44
79	Amusement and recreation services	4,773	955,592	0.81
80	Health services	50,537	8,429,654	0.97
81	Legal services	4,856	892,143	0.88
82	Education services	7,441	1,703,729	0.71
83	Social services	9,395	1,630,119	0.94
84	Museums, botanical, zoological garden	744	60,317	2.00
86	Membership organizations	11,055	1,837,318	0.98
87	Engineering and management services	16,709	2,384,483	1.14
89	Miscellaneous services	1,163	179,235	1.05
	Nonclassifiable establishments	3,944	856,953	0.75
	State and local government	108,800	14,771,000	1.20
	TOTAL	652,653	106,058,220	

Source: Private-sector employment data are from County Business Patterns; state and local government employment data are from the Ohio Bureau of Employment Services.

from the creation of new basic-sector jobs. The adjustments—made for mining, eight manufacturing industries, and selected services sectors that have predominantly nonlocal clients or cater to visitors—are listed in the last two columns of table 3.2. The most significant adjustment increased basic employment in the state and local government sector by more than 30,000, to account for the influence of Columbus's two "anchors"—the state capital and Ohio State University. The adjusted multiplier for Columbus is a significantly lower and more reasonable figure: 4.90. This multiplier assumes an average propensity to spend locally of 0.796 in equation (3).

TABLE 3.2

Estimated 1989 Basic/Nonbasic Employment,
Columbus Metropolitan Statistical Area

		Unadjusted		Adjusted	
SIC	*Industry*	*Basic*	*Non-basic*	*Basic*	*Non-basic*
	AGRICULTURE	666	3,011	666	3,011
	MINING	0	978	400	578
	CONTRACT CONSTRUCTION	0	28,040	0	28,040
20	Food and kindred products	0	8,266	0	8,266
21	Tobacco products	0	0	0	0
22	Textile mill products	0	235	0	235
23	Apparel and other textile products	0	515	0	515
24	Lumber and wood products	0	111	0	111
25	Furniture and fixtures	0	1,949	0	1,949
26	Paper and allied products	0	1,829	0	1,829
27	Printing and publishing	0	8,789	0	8,789
28	Chemicals and allied products	0	4,483	2,000	2,483
29	Petroleum and coal products	0	150	0	150
30	Rubber and miscellaneous plastic products	829	5,562	4,000	2,391
31	Leather and leather products	0	375	0	375
32	Stone, clay, and glass products	3,930	3,245	5,000	2,175
33	Primary metal industries	0	2,645	1,500	1,145
34	Fabricated metal products	0	8,171	3,000	5,171
35	Machinery, except electrical	0	8,907	3,000	5,907
36	Electric and electronic equipment	0	7,304	3,000	4,304
37	Transportation equipment	0	3,215	0	3,215
38	Instruments and related products	0	4,719	2,500	2,219
39	Miscellaneous manufacturing industries	0	1,224	0	1,224
—	Administration and auxiliary		5,562	0	5,562

Industry Targeting

The Columbus Area Chamber of Commerce has in place a sophisticated program of industrial recruitment and promotion, which, along with the area's major utilities, has been active in this arena for many years. The Chamber produces a great quantity of statistical and marketing information on Columbus; it contacts suspects and prospects on a regular basis, hosts site visits of site-selection consultants and corporate officers, and facilitates the location process for companies that have selected Columbus.

Columbus developers are applying economic base theory in two ways. First, industrial recruitment and promotional efforts clearly are aimed at ex-

TABLE 3.2 (continued)

Estimated 1989 Basic/Nonbasic Employment, Columbus Metropolitan Statistical Area

		Unadjusted		Adjusted	
SIC	*Industry*	*Basic*	*Non-basic*	*Basic*	*Non-basic*
	TRANSPORTATION AND UTILITIES	0	32,530	0	32,530
	WHOLESALE TRADE	1,901	37,865	1,901	37,865
	RETAIL TRADE	9,696	118,983	9,696	118,983
	FIRE	13,448	41,855	13,448	41,855
70	Hotels and other lodging places	0	6,650	6,500	150
72	Personal services	223	7,028	223	7,028
73	Business services	10,083	29,226	15,000	24,309
75	Auto repair, services, and garages	0	4,669	0	4,669
76	Miscellaneous repair services	141	2,293	141	2,293
78	Motion pictures	0	935	800	135
79	Amusement and recreation services	0	4,773	0	4,773
80	Health services	0	50,537	0	50,537
81	Legal services	0	4,856	2,000	2,856
82	Education services	0	7,441	2,000	5,441
83	Social services	0	9,395	0	9,395
84	Museums, botanical, zoological garden	373	371	600	144
86	Membership organizations	0	11,055	0	11,055
87	Engineering and management services	2,036	14,673	7,337	9,373
89	Miscellaneous services	60	1,103	60	1,103
	Nonclassifiable establishments	0	3,944	0	3,944
	State and local government	17,903	90,897	48,505	60,295
	TOTAL	61,289	591,364	133,277	518,376

panding the basic sector. The Chamber seeks national and international companies that produce for markets extending well beyond the Columbus area. Although these efforts are targeted in various ways, they do not necessarily focus on industrial sectors that would generate strong linkages and high multiplier effects. Instead, attention is paid to companies that want to make sizable investments and locate large numbers of jobs in Columbus. Second, job announcements associated with the locations of new facilities often include estimates of overall job creation and use multiplier models to generate these results. The Columbus economic base multiplier estimate given above, and the Columbus input-output multipliers in Appendix 3.2, could be used for such an impact analysis.

Economic developers in Columbus engage in activities that support new business development and existing businesses and industries, but these activities are not pursued for the purpose of export diversification. On the other hand, most Columbus developers are keenly aware of the need for efficient public infrastructure and services as a means of attracting basic-sector companies to the region. Development publications extol the availability and quality of infrastructure for companies and for their employees and families. Through continuing annexation, the city of Columbus has managed to retain its share of the metropolitan tax base much more effectively than many other large central cities. Since the 1950s, the geographic area of the city of Columbus has quadrupled from about 50 square miles to nearly 200. Due primarily to its aggressive annexation policy, Columbus was the fastest-growing large city outside the Sunbelt during the 1980s. The annexations were politically feasible because suburban school districts have not been consolidated with city schools. Further discussion of these issues is available in Cox (1988). The German cultural heritage of the region reinforces the goal of effective local government. Developers also seem interested in ensuring that capacity constraints do not constrain growth, and the people of Columbus seem willing to commit public capital to needed infrastructure projects.

More widespread or subtle applications of economic base theory are not apparent in Columbus-area economic development practice. Yet the recruitment strategy, which is supported by economic base theory, is a pillar of local practice. Economic base theory provides a common language for local advocates of economic growth.

III. Elaboration and Criticism

First, economic base theory confuses two distinct objectives: maintenance of a favorable, regional balance of trade (commodity exports being worth more than commodity imports) and a focus on "critical" industries (those industries that are vulnerable to outside competition).[7] In other words, the model does not distinguish clearly a region's balance of trade position from its competitive position; the developer needs to understand both clearly. The former deals with all current flows entering or leaving the region. The latter assesses the ability of firms in the area to market their traded products (exports). With regard to the first objective, regions must export in order to bring in money to pay for needed imports; export base theory advises practitioners to focus on basic industries, to ensure an adequate level of export earnings. However, to be most effective, any emphasis on balance-of-trade should necessarily treat the import side of the ledger, as well as the export side. Developers also should focus on reducing imports, which requires paying attention to the fortunes of local serving (nonbasic) industries. Suddenly the priorities of economic base theory are not so clear: the developer must ensure the healthy growth of both basic and nonbasic sectors—which, together, include every firm in the region!

The logic of economic base theory is not carried to this extreme, however, because of the second, critical-industries objective. This objective has motivated, and confused, thinking about the economic base. Economic base theory properly emphasizes those industries that, if lost, would constitute the greatest loss to the community. Moreover, through industrial recruitment, it identifies those industries, which, if gained, would constitute the greatest benefit to the community. However, every critical industry is not necessarily in the basic sector.[8]

A second problem with economic base theory is the arbitrary nature of dividing the regional economy into basic and nonbasic sectors. The issue ultimately is conceptual rather than empirical, although there are plenty of reasons why the empirics of bifurcation are likely to fail, including lack of explicit regional export data, use of employment instead of output information, and financial and operational infeasibility of census surveys.[9] Ultimately, the difference between basic and nonbasic activity is a function of the organization of industry, not the fundamental economic growth dynamic identified by the theory. If a local exporting firm, for example, buys $200 worth of intermediate inputs from local suppliers, then exports the product at a price of $500, only the value added to the product by the firm ($300) is considered an export. If the firm were fully vertically integrated, thus supplying its own inputs, the entire value added ($500) would be considered an export. In the first case, the input suppliers sold their product to a local industry. In the second, they sold directly to the exporting industry, and no intermediate sales of any part of the product would be considered nonbasic; the basic-nonbasic distinction begins to lose its meaning.[10] The distinction becomes more problematic as the local economy under consideration becomes larger, because the specialization and differentiation of the local economy are likely to increase as the metropolitan area grows. In other words, the multiplier is sensitive to the size of the region. Moreover, by reducing the economy to two sectors, the two-sector model suppresses both interindustry relations and the multiplier effects of different export industries. Input–output analysis can overcome some of the difficulties. Indeed, given the greater availability of reasonably priced input–output data at the county level, there is little justification for using the simple base multiplier (see Appendix 3.2).

A third, and related, fundamental problem with economic base theory is that, by focusing on exports and the basic sector, the theory neglects the reality of the regional economy as an "integrated whole of mutually interdependent activities" (Blumenfeld 1955, p. 121). The theory ignores autonomous investment and the role of the nonbasic sector in stimulating economic growth.[11] Since all export firms require local services to produce, all local services are basic, in this sense, and constitute the foundation upon which further regional growth can take place. According to Blumenfeld:

> It is thus the "secondary," "nonbasic" industries, both business and personal services, as well as "ancillary" manufacturing, which constitute the real and lasting strength of the metropolitan economy. As long as they continue to function efficiently, the metropolis will always be able to substitute new "export" industries for any which may be destroyed by the vicissitudes of economic life." (1955, p. 131)

The true economic base of the city over the long term consists of a large local market, the availability of skilled and talented labor, and an array of business services, including public infrastructure. It is business services and other "secondary" industries that, together with the availability of labor of all kinds, enable the metropolis to sustain, expand, and replace its "primary" industries. Local industries are more permanent and stable than export industries. While the existence of a sufficient number of export industries is indispensable for the continued existence of the metropolis, each individual "export" industry is expendable and replaceable. Blumenfeld's arguments, which constitute a reversal of the causal direction of the original model, were picked up years later by Wilbur Thompson and others (see chapter 11).[12]

The criticisms related to size of region and reversal of sectoral importance raise the relationship between specialization and size. The economic base model would appear to work best in smaller cities or "one industry" towns where the division of labor is severely restricted by small market size. External demand would surely enhance the prospects of local economic growth. At the other extreme, large metropolitan areas are inherently more diverse by virtue of their ability to support multiple specializations. These export industries may well be less important than "secondary" industries. A large local market also offers a more generous source of demand for local production.

The economic base model can be used as a tool for near-term predictions of economic growth and for impact analysis as long as (1) the industrial composition of the basic sector does not experience substantial change, (2) nonbasic industries retain their competitive position in the local economy, and (3) the demand for export products is the factor causing local change. The model has limited applications for long-term analysis of regional economic development precisely because these conditions are not likely to remain the constant for long.

At the same time, the model's equation of growth with increases in welfare deserves close scrutiny. Economic growth may benefit some members of the community while imposing costs on others. Furthermore, the model implies that monopolies in the nonbasic sector, be they private or public, can be ignored. Because these monopolies cannot be replaced, their activities should not be ignored in order to protect consumer welfare. Conversely, companies in the basic sector have to compete for market share. From the company's perspective, their sales are of concern. But, from the perspective of the local consumer, the availability of inexpensive goods, regardless of source, is the central issue. The point here has to do with the objective of economic development. Is the objective employment growth or increased consumer welfare?[13]

In a similar vein, economic base theory can be used to anticipate the local interests that would favor and benefit from growth or that would oppose it. Firms in the local sector that enjoyed economies of scale would benefit. Owners and managers of these firms would tend to be growth boosters because, for their returns, bigger would be better. Such firms may well include local newspaper companies, communications firms, electric/gas/water utilities, banks and other financial institutions, and possibly construction contractors and companies providing business services. Further, the value of urban land should increase as growth occurs. Thus, landowners—including farmers with property on the urban fringe, real estate firms, and other transactions-oriented firms whose commissions are tied to asset values—should expect to realize capital gains as values increase. On the other hand, growth could create diseconomies in the form of congestion, crime, higher tax rates, higher rents, and/or higher prices for local commodities. Local residents who would experience these diseconomies may well oppose continued growth. However, residents who got better jobs, or who earned more, would be less likely to oppose growth. Since economic base theory helps identify the expected costs and benefits of growth, it also helps developers understand pro- and anti-growth forces found in most U.S. localities.

Discussion Questions

1. How would you gauge your local economy's competitive advantage from the perspective of economic base theory? From Blumenfeld's perspective?
2. Do you think your local economy has a favorable balance of trade (exports greater than imports)? Does its balance of trade position matter? How has it changed over time?
3. Sectors that export or serve local consumers are relatively easy to classify. What about local sectors whose output is primarily purchased by an export sector?

4. Why may some local sectors with very low location quotients be critical industries? (See note 8 to this chapter.)
5. In which direction do regional (input-output or economic base) multipliers change when:
   (1) the region grows larger?
   (2) the regional economy becomes more interdependent?
   (3) all final demand sectors except exports are treated as endogenous?
6. Why is the economic base model potentially more useful for understanding contractions rather than expansions?
7. Why is economic base theory more useful for analyzing economic growth than for understanding economic development?
8. How might you get an economic development advisory board that accepted an economic base view of the world interested in long-term development issues?

Appendix 3.1

DETERMINING REGIONAL EXPORT ACTIVITY

In order to calculate the economic base multiplier, regional economic activity must be divided into activity that is basic (serving the export market) and activity that is nonbasic (serving the local market). The fundamental problem facing the practitioner interested in conducting an economic base analysis is lack of available data on regional imports and exports. Although a survey of area employers obviously represents the best possible bifurcation method, such an approach usually is prohibitively expensive.[14] As a result, several somewhat arbitrary, but relatively simple and inexpensive, methods have been developed to determine the level of regional export activity. The most commonly used approaches are the assignment method, an approach based on location quotients, and the minimum requirements technique.[15]

The Assumption or Assignment Method

The simplest way to divide the regional economy is to determine, on the basis of expert judgment, which sectors are basic and which are not. Of course, few regional industries will produce entirely for export (or to satisfy only local demand). This difficulty might be assumed away with the hope that overestimates of export activity in some industries will be canceled out by underestimates of export activity in other industries. Or, supplementary information (for example, location quotients and surveys) might be used to determine the proportion of export activity in mixed industries.

The Use of Location Quotients

One of the simplest, most commonly used methods of determining regional export activity involves location quotients. The location quotient is a measure of the sectoral specialization of a given region relative to a reference region, usually the nation.[16] Using employment data, location quotients for each regional industry are calculated as follows:

$$LQ_{ir} = \frac{E_{ir}/E_r}{E_{in}/E_n} \tag{4}$$

where i denotes the industry, r the region under study, n the nation, and E employment. Although any relevant measure of economic activity can be used to calculate location quotients, regional data constraints frequently make employment the variable of choice.[17] As we demonstrate below, the use of employment as a proxy for output requires the use of strict, not particularly realistic, assumptions.

From equation (4), if regional employment in industry i, as a proportion of total regional employment, is .2, while national employment in the same industry as a proportion of total national employment is .1, the location quotient for industry i will be 2 (.2/.1). It is easy to see that, when the proportion of regional employment in a given industry exceeds the proportion of employment in the same industry nationwide, the location quotient will exceed 1. When this is the case, the regional economy is said to be more specialized in the particular industry than the nation as a whole. Similarly, when the location quotient is less than 1, the regional economy is considered less specialized in the given industry than the nation. A location quotient equal to 1 indicates that the region and the nation are equally specialized in the study sector.

The basic premise underlying use of the location quotient to calculate regional export activity per industry is that the amount by which any regional industry's share in total regional production exceeds that national industry's share in total national production must be devoted to export. Alternatively, it is assumed that any regional industry's proportion of regional production used to satisfy local demand must be equivalent to that national industry's proportion for the nation as a whole. Any production in excess of this amount must be serving extraregional demand. Therefore, when $LQ_{ir} > 1$, regional export employment in industry i may be estimated by

$$X_{ir} = \left[1 - \frac{1}{LQ_{ir}}\right] E_{ir} \quad (5)$$

Therefore, total export activity, or the basic sector of the regional economy, is

$$X_r = \sum_i X_{ir} \quad (6)$$

By substituting equation (4) for LQ_{ir}, rearranging, and factoring out E_{in}, we can derive a more theoretically useful expression for exports (Isserman 1980):

$$X_{ir} = \left[\frac{E_{ir}}{E_{in}} - \frac{E_r}{E_n}\right] E_{in}. \quad (7)$$

TABLE 3.3

Two-Region, Two-Industry National Economy

	Region A Employment	*Region B Employment*	*National Employment*
Industry	100	100	200
Industry	500	900	1,400
Total	600	1,000	1,600

Consider a two-region, two-industry national economy as indicated in table 3.3. Employment in industry *i* in region A represents one-half of total national employment in industry *i*. Total employment in the region, however, represents less than one-half of total national employment. Given these figures, what can be said about export employment in industry *i* in region A?

From equation (7), it is clear that the location quotient approach to the determination of regional export employment rests on two critical assumptions (Isserman 1980, pp. 157–58). First, if productivity per employee in industry *i* is equal in the region and the nation, then the region's share of national industry *i* employment (E_{ir}/E_{in}) indicates its share of output in industry *i*. Using this assumption in the above example, we can conclude that region A produces one-half the nation's output of industry *i* (100/200 = .5). Second, if per-employee consumption of industry *i* output is the same in the region and the nation, then the region's share of total national employment (E_r/E_n) represents its share of consumption (600/1,600 = .375). Given these assumptions, since region A produces exactly half the national industry *i* output, but contains less than half the total national employment (and therefore consumes less than half the total national output), it cannot be consuming all its industry i output. Carrying out the calculations using equation (7), 25 out of 100 industry i employees in region A produce for export.

Two additional theoretical implications follow from the two primary assumptions. First, there is no *cross-hauling:* the region cannot import, for its own consumption, the same product that it exports. This is clear from equation (7), since a region's share of consumption is subtracted entirely from its own share of output of industry i. Second, there are no net national exports or imports of industry i. If it is assumed that the region's share of total national employment represents its share of consumption of industry i, then the nation cannot be exporting or importing i. Following the example given in table 3.3, region A satisfies its own demand for i while also exporting part of its output of i to region B. Likewise, region B is able to fully satisfy its internal demand for j while also exporting to region A.

A major problem with the location quotient approach is that, if cross-hauling does in fact occur, the method estimates net, rather than gross, exports. If region A imported part of its consumption of i, more of its internal production would be available for exports than the location quotient method would indicate. Clearly, the assumption of no simultaneous importing and exporting of the output of a given industry is invalid, since it represents the aggregation of numerous different types (and brands) of products, some of which may be imported, exported, or both. Although the underestimation of exports can be reduced by using highly disaggregated industrial data, the problem can never be completely eliminated.[18]

The Minimum Requirements Technique

The minimum requirements technique also is a popular method of estimating regional export activity.[19] The method is nearly identical to the location quotient approach, except that the regional economy is compared to other similarly sized regions rather than to the nation. Application of the technique involves three steps: (1) assembly of a sample of regions of comparable size; (2) for each industry i, identification of the region that has the minimum share of total regional employment in industry i; and (3) estimation of regional export activity for each industry according to the following:

$$X_{ir} = \left(\frac{E_{ir}}{E_r} - \frac{E_{im}}{E_m} \right) E_r \tag{8}$$

where m denotes the region with the minimum share of employment in industry i. As equation (8) indicates, all other nonminimum share regions are assumed to export the difference between industry *i*'s share of employment in the region and the industry's share in the minimum region, expressed in employment terms.

The following expression, similar to equation (7) above, can be derived from equation (8) by collecting terms within the bracket and multiplying and dividing by E_{in} :

$$X_{ir} = \left[\frac{E_{ir}}{E_{in}} - \left(\frac{E_{im}}{E_{in}}\right)\left(\frac{E_r}{E_m}\right)\right]E_{in} \quad (9)$$

Isserman shows that the first term inside the bracket (E_{ir}/E_{in}) represents a proxy for regional output given the equal productivity assumption. The consumption term, however, indicates that the region of study is assumed to consume a regional output share identical to the minimum region's share of national *i* output, scaled by the relative size of the study and minimum regions (E_r/E_m). As noted, this treatment of consumption is the fundamental difference between the minimum requirements approach and the location quotient approach.[20]

To reiterate, the minimum requirements approach first assumes that all production of *i* in the minimum region is for local consumption. Then, consumption per employee in the study region from *local production* of *i* is assumed to be equivalent to consumption per employee in the minimum region (which, by definition, neither imports nor exports the products of industry *i*). The third assumption also applies to the location quotient approach: equal productivity per employee in the region and the nation.

Other Bifurcation Techniques

The three methods of dividing the regional economy discussed above are the simplest and most common of a wide variety of alternative approaches. In order to relax three of the assumptions underlying the location quotient approach, Isserman (1977), in equation (7), proposes using regional/national labor productivity and consumption ratios by industry, as well as the ratio of net national exports of industry *i* to national output of *i* (Isserman 1977). Richardson (1985) briefly reviews this method, as well as Norcliffe's (1983) consumption-based location quotient method, Moore's (1975) simple regression application of the minimum requirements technique, and the Mathur and Rosen (1972, 1974, 1975) regression method. Finally, Stabler and St. Louis (1990) describe a way to use input-output analysis to classify regional economic activity. Their method also attempts to distinguish between direct and embodied exports.[21]

Appendix 3.2

INPUT–OUTPUT MULTIPLIERS

The rather arbitrary splitting of the economy into two sectors can be avoided by using an input-output model. With the availability of input-output multipliers through the Regional Economic Analysis Division of the Bureau of Economic Analysis, U.S. Department of Commerce, developers now have sectoral information and industry-specific multiplier estimates for states and metropolitan areas. Similarly, many empirical problems with economic base models can be overcome by substituting input-output models and using the latter for multiplier analysis and prediction. Billings (1969) and Garnick (1970) demonstrate the mathematical identity between the economic base multiplier and an aggregate (weighted) input-output multiplier when the final demand sectors of the input-output model are identical to the exogenous sectors of the economic base model.

The input-output equations shown below in matrix form are comparable to the economic base formulation shown in equations (1)-(3) in the text:

$$\mathbf{AX} + \mathbf{Y} = \mathbf{X} \tag{10}$$

$$\mathbf{Y} = \mathbf{X} - \mathbf{AX} \tag{11}$$

$$\mathbf{Y} = (\mathbf{I} - \mathbf{A})\,\mathbf{X} \tag{12}$$

$$(\mathbf{I} - \mathbf{A})^{-1}\,\mathbf{Y} = \mathbf{X} \tag{13}$$

where **I** is an identity matrix, **X** is a vector of total output, **Y** denotes a vector of final demand, **A** is a matrix of national technological coefficients or more usually trade coefficients in regional tables, **AX** is a matrix of intermediate demand, **(I–A)** is called the Leontief matrix, and $(\mathbf{I}-\mathbf{A})^{-1}$ is the matrix of interindustry multipliers or the Leontief inverse.

Equation (10) is an accounting identity which says that output used by industries to produce commodities in the current period (**AX**) plus output used by households, governments, and businesses as capital investment (**Y**) equals total output (**X**). Final demand (**Y**) includes sales to local entities as well as exports to entities located outside the region. The table of coefficients (**A**) shows the value of inputs from any industry needed to produce a dollar's worth of output from each industry. In national input-output tables, these interindustry coefficients are estimated so as to reflect the average technology in each industry. In other words, certain amounts of input from other industries, plus certain amounts of labor, are needed to produce industry outputs. In regional input-output tables, interindustry coefficients usually are estimated as trade coefficients. In other words, inputs from local industries per dollar of output are shown in the **A** matrix; imports, which are supplied by any nonlocal source, and labor are treated as separate inputs.

Equation (13) connects final demand (**Y**), premultiplied by the Leontief inverse matrix, to total output (**X**). The equation says that, to satisfy final demand by industry (**Y**), the economy must produce a larger amount of total output (**X**). The interindustry multiplier effects contained in the Leontief inverse matrix determine the industry-specific ratios of total output to final demand.

The fact that the Leontief inverse matrix contains multipliers can be grasped by recognizing that one can invert a matrix by solving a power series. In this case, the power series contains the unit amount of final demand (**I**), plus the first round of inputs needed to produce one unit of output for all industries (**A**), plus the second-round effects inputs needed to produce the first-round inputs (**A*A**), plus the third-round inputs needed to satisfy second-round inputs (**A*A*A**), and so on. The consecutive rounds of inputs and outputs are represented by the sum of the power series and are equal to the multipliers for each industry.

In most input–output applications, the impact of changes in final demand are analyzed. Therefore, one can modify equation (13) to relate a change in **Y** to a change in **X**. The Leontief inverse matrix is not changed; as with the economic base multiplier, the interindustry multipliers are assumed to remain the same. With this input–output model, the economic developer can see changes in each industry that stem from a change in the final demand of a particular industry or ascertain the overall effects on output of projected changes in final demand. Thus, the input–output model can be used for near-term forecasting or for impact analysis.

Input–Output Multipliers for Columbus, Ohio

Professor Randall W. Jackson, of the Geography Department at Ohio State University, provided a seven-sector input–output model for the Columbus metropolitan area that can be used to illustrate estimation of interindustry multiplier effects. This model presents a more disaggregated view of the regional economy, compared to the two-sector economic base model.

Let us assume, consistent with economic base theory, that an economic developer wants to attract some manufacturing activity to Columbus and to estimate the resulting impacts on output, income, and employment there. Unlike the economic base model that is biased by the procedure selected for calibrating the economic base multiplier, this input–output model yields replicable results for the Columbus region, given the same exogenous change in final demand.

TABLE 3.4

Input-Output Impacts of a $10 Million Increase in Final Demand in Manufacturing in Columbus, by Industry

Sector	*Impacts*
Agriculture/Mining	$691,699
Construction	$178,471
Manufacturing	$5,831,752
Transportation/Trade	$994,053
Services/Government	$1,441,961
Total Intermediate Output	$9,137,937
Households	$5,697,114
Imports	$5,072,222
Other Value-Added	$2,086,699
Total Value-Added	$12,856,035
Total Output	$21,993,973
Employment	155

Table 3.4 shows the impacts of increasing final demand for products from the manufacturing sector by $10 million. Value-added increases by almost $13 million while intermediate output increases by over $9 million. Because output in the high-wage manufacturing sector increases more than any other sector, average income per job is rather high; the 155 new jobs receive almost $37,000 each.

Appendix 3.3

Regional Economic Accounting

Five regional economic accounts are used to track the economic flows and stocks that measure the process, structure, and outcomes of regional growth and development. These accounts are modeled after the national "social accounts" produced by various federal agencies. The Bureau of Economic Analysis in the U.S. Department of Commerce produces most of the available subnational information.[22]

By no means does a complete set of regional accounts exist in the United States or any other country. Nonetheless, there are two reasons to cover the logic and measures in the accounts. First, some regional measures are published and are widely used as measures of the structure and performance of regional economies. Second, the measures themselves, whether available or not, are indicators of important regional concepts and relationships. Understanding these indicators increases our understanding of the concepts they represent.

The best-known measures of regional economic activity are gross domestic product (GDP), regional income, and personal income. Personal income is often measured in per capita terms, while GDP is often measured per employee. Per capita income is the most common measure of economic well-being. The distribution of personal income can be used as a measure of social inequality. These measures are derived from regional income and product accounts.[23] GDP is the value added of all production taking place within the region. Usually it is calculated by estimating and summing the incremental contribution of each industry in the region. Regional income is the sum of all income earned by regional owners of productive factors (land, labor, and capital). Personal income includes both earned income from labor and property as well as transfer payments to regional households. The major uses of personal income are consumption, tax payments, and saving.

Input-output and money flow accounts are more comprehensive flow accounts. The former account for all current flows among industrial sectors and "final demand" sectors. These flows include primary inputs to sectors, which generate factor income, and sectoral output to final demand, which represent uses of income by households, businesses, governments, and rest-of-the-world (ROW). The latter are, in fact, the components of income made famous by Keynes: consumption, investment, government spending, and exports. Input-output accounts focus on the remaining flows, namely the intermediate flows (inputs and outputs from one sector to another). Thus, input-output accounts recognize intermediate sales and purchases in addition to the value added of production and factor income.

Input-output accounts are transformed into the input-output model discussed in Appendix 3.2. The most important calculation is to divide inputs from each sector by the output of that sector. In national tables, these fractions or coefficients are called technical coefficients because they portray the production function for each sector. Together they show the structure of production (that is, production relationships) for the entire economy.

In regional tables, input-output relationships usually reflect trade (sales and purchases) between sectors. The computed coefficients are known as trade coefficients. This important distinction is the main difference between regional input-output tables and the national table. As noted in this chapter, national technical coefficients are used to identify groupings or clusters of industries that interrelate heavily. To the extent that these clusters locate in the same region, they form industrial complexes. Complexes are less complete than clusters. To produce output, industries in the complex must import commodities. A technical coefficient, then, equals the trade coefficient for the sector, plus the import coefficient (the value of imports to that sector per dollar of sectoral output) for that sector.[24]

Money flow accounts are more comprehensive than any flow account because they include all monetary transactions, specifically the flows of cash and credit between primary sectors (households, businesses, and governments) and financial intermediaries. Some of these transactions relate to annual income and production and to the intermediate sales and purchases in input-output accounts. They also include all other transactions occurring during one year (for example, the production and sale of new housing, as well as the sale of existing houses). Essentially, these transactions change the claims on existing assets. Although the accounts are very important at the national level, in terms of setting and monitoring monetary policy, they rarely exist at the regional level. They are useful, however, in showing how sales and purchases are financed by cash and credit in the economy. In other words, money flow accounts keep track of the money sphere, while the aforementioned accounts focus on the "real" sphere.

Balance of payments accounts also are typically unavailable for regional economies. Yet, conceptually, they are extremely useful in terms of understanding the regional economy as an open region trading with the rest of the world, since they provide a systematic annual record of all transactions between units resident in a region and "foreign" (that is, nonlocal) units over one year. The three subaccounts—current account, capital account, and cash/currency account—form a single double-entry system drawn from the perspective of the region. The current account includes all transactions that give rise to or use up regional income. The capital account records all changes in claims (ownership) between the region and the ROW. The cash/currency account shows only net currency movements.

Credits on current account or capital account are transactions that result in inflows of funds to the region. Conversely, debits reflect outflows of funds. The equal and opposite monetary side of the transaction is recorded in the cash/currency account. Although, in theory, an overall balance should exist, errors and omissions are included for final reconciliation. Balance of payments are an important piece of the international trade picture. In the long run, trade between countries can proceed only in a fashion such that balance of payments in each trading country are reconciled. The information described here is rarely available for subnational regions, in part because only one national currency is used in interregional transactions. The most important relationship shown is referred to as the balance of trade, which is the difference between exports and imports. Often, the balance of trade focuses exclusively on merchandise exports and imports. In other words, only the trade of goods is considered.

TABLE 3.5

Hypothetical Community Balance of Payments Accounts (in millions)

Item	*Credit*	*Debit*	*Net*
1. Current Amount			
Merchandise and service exports	$22		
Merchandise and service imports		$51	($29)
Wages and salaries of out-commuters	$57		
Wages and salaries of in-commuters		$41	$16
Property and proprietary income to residents	$2		
Property and proprietary income to foreigners		$16	($14)
Transfers received by residents	$30		
Taxes paid by residents		$40	($10)
Balance on current account			($37)
2. Capital Account			
Grants and loans to locals	$17		
Mortgages extended to locals	$4		
Credit extensions to rest-of-the-world		$5	$16
Direct foreign investment	$9		
Direct investment by locals		$0	$9
Balance on capital account			$25
3. Cash Account			
Net currency movements			$2
4. Errors and Omissions, Net			$10

A simplified set of accounts is shown in table 3.5 for a *hypothetical* U.S. central-city area. In principle, a region should sell enough commodities to the ROW to finance imports from the ROW. This hypothetical community has a negative balance of trade. The accounts also compare labor and property income. The wages and salaries of residents who commute outside the region may be compared to the wages and salaries of in-commuters. These flows should be minimal to the extent that the regional economy represents an entire labor market area (that is, one commuter shed; see Appendix 4.1). In this case, net inflows result from commutation. More important are the flows of rents, dividends, and interest. A comparison of inflows to outflows indicates the claims residents have on foreign tangible assets and foreigners have on local tangible assets. Foreigners receive far more property income from the community than local residents receive from other areas. Finally, the current account shows both taxes paid to external governments and transfers received from them. The community pays more in taxes than it receives in transfers.

In addition to grants and loans, the capital account shows foreign investment in the region compared to investment by residents outside the region. It is important to note that foreign investment is a credit because it brings funds into the region. In the hypothetical community, foreign investment, loans, and grants are the prominent flows. The overall balance of payments is found after estimating errors and omissions. In this case, the deficit on current account is financed largely from net inflows on capital account.

Two insights from these accounts should be mentioned. First, the balance of trade and balance of payments relationships are worth considering over the long term. These relationships help us understand the role the region has played in the global economy and how that role is changing over time. Second, the relationships in the accounts underscore the need to build viable local industries. As the economic base of a region becomes less competitive, exports decline and the balance of trade becomes more negative. Foreign investment, grants, loans, or transfers can finance imports, but they are not a long-term solution. Foreign investment gives rise to future payments of property income to foreign owners of local assets (called "repatriated profits" in the international trade literature). Other inflows either have to be repaid or they can dry up.[25]

Wealth accounts are neither available at the regional level nor mentioned in regional development theory; yet they provide important insights into the regional economy. These accounts are comparable to business balance sheets, in that they account for economic stocks rather than flows. Thus, they record the types and value of productive assets in the region and show the structure of ownership (liabilities and net worth) at one point in time.

Wealth accounts by type of assets would include all natural resources of the region, as well as built assets. They would include both private fixed assets and public infrastructure and facilities. By tracking changes in the accounts from year to year, one can see how resources and other tangible assets are being depleted and depreciated or increased. The accounts also show the ownership and control of regional wealth. All foreign owners of local assets are considered, in that the value of assets they own is recognized as liabilities. The value of locally owned assets is equity or net worth.

Wealth accounts can also be used to estimate the regional distribution of wealth. Empirical evidence indicates that the wealth distribution in the United States is much more unequal than the income distribution. These estimates can be used to monitor changes in the wealth distribution over time.

Another application of wealth accounts is that related to understanding power relations. Information in these accounts can help the economic developer gain an overall picture of the type, structure, and distribution of regional wealth. With this information, the developer can gauge the wealth effects of economic growth.

Notes

1. Hoyt and Weimer (1939) developed the economic base model as a means of estimating the prospects of local economies. Their work was used to gauge the economic risk of purchasing residential mortgage loans from various markets in the period when secondary markets were being established in the United States.

In "Homer Hoyt on the Concept of the Economic Base" (1954), Hoyt outlines the microeconomic foundations of the theory. Because industries are subject to price competition, production costs in one metropolitan area, compared to others, will determine the future trends in basic employment. Furthermore, exports are needed to pay for imports. Hoyt notes that the base-service ratio will vary in one city over time as well as by city size and level of income. Later, in Hoyt (1961), he takes issue with Blumenfeld's (1955) critique. Regardless of other influences on growth, he argues that his empirical analysis suggests a "causal" relationship between growth of the economic base and overall growth. Therefore, urban planners can use the model to forecast employment and population growth: "In short, it has not yet been demonstrated that an urban region can grow substantially in population by an increase in its non-basic industries only. It may reduce its reliance on imports by greater diversification, but a new impetus to growth must come primarily from basic employment" (Hoyt 1961, p. 56).

2. The derivation of (3) follows:

$$b = n - an = (1 - a)n, \left(\frac{1}{1-a}\right)b = n.$$

The model can also be estimated, using *changes* in employment, income, or output. In this instance, the multiplier estimates the marginal impacts and marginal propensity to spend locally, rather than the average values.

3. Another way of putting this is that supply is assumed perfectly elastic at fixed prices. Elasticity of supply (explained in chapter 4) describes the degree to which the quantity supplied of a given commodity increases or decreases with changes in its price. Perfect or near-perfect elasticity implies that producers are extremely responsive to small price changes; capacity constraints do not limit production and therefore do not lead to price increases.

4. Development practitioners are not the only ones who continue to find economic base theory useful, since it continues to generate a fair amount of research. See, for example, LeSage (1990) and LeSage and Reed (1989) on the short- and long-run utility of the theory; an application of LeSage and Reed's methodology by Kraybill and Dorfman (1992); research on the derivation of multipliers in the presence of local income leakages (Frey 1989, Farness 1989); and work on a new input-output-based methodology for classifying regional economic activity into basic and nonbasic sectors (Stabler and St. Louis 1990).

5. The three strategies for export expansion described in this paragraph are illustrated in two issues of *Economic Development Review: The Journal for the Economic Development Practitioner*. The winter 1991 and summer 1996 issues focus on existing industry development; the spring 1991 issue addresses the recruitment of investment. One article in the latter issue explicitly relates economic base theory

to the targeting and recruitment of economic activity. See Miller, Gibson, and Wright (1991). New business development in a rural context is considered in the fall 1992 issue.

6. Import substitution can be developmental, provided the city is large enough to support enterprises at an efficient scale of operation. Import enhancement through product improvements is even more important as growth proceeds and the division of labor deepens (Jacobs 1985).

7. See Blumenfeld (1955), who presents a readable description of the model and the most trenchant critique of its many weaknesses. The tenacity of economic base theory should be admired when one considers that his criticisms were published more than forty years ago. He argues that the model works best when considering a small city with a clearly defined export sector and a totally dependent service sector. Blumenfeld notes the theory's physiocratic and mercantile overtones as well. The former relate to the primacy of agriculture, mining, and manufacturing over other sectors and to the idea that all wealth originates in the basic sector. The mercantile overtones are more significant. Because economic base theory reduces wealth creation to the growth of export earnings, the important service-provision role of the city is trivialized. Money, however, is not the wealth of nations, as Adam Smith pointed out; rather, the expansion of productive capacity that leads to greater competitiveness results in real national wealth. Mercantile thinking as reflected in the economic base model may be attractive to economic developers because it describes how a particular place can make money. The pro-business bias of the model becomes clear when we consider its implications for consumer welfare.

8. Blumenfeld, in a somewhat apocalyptic example, illustrates the rationale: "If General Motors closes shop at Flint, no efforts to promote the development of department stores will save the town. On the other hand, if a Flint department store closes down, but the General Motors plant continues to operate as before, it will soon be replaced by other stores. Therefore, the thing to worry about, the 'base' of the Flint economy, is the automobile industry; once that works, the 'services' will take care of themselves" (1955, p. 121).

The difference between the department store and the automobile plant is that the latter must compete with firms outside the region and country, whereas the former competes only with those within the community. The definition of a "critical" industry, then, is not whether it exports, but whether its area of potential competition extends beyond the region. Even if available bifurcation methods can measure basic activities accurately (see Appendix 3.1), a "criticality" approach to economic base theory suggests that efforts to refine these methods to measure export activity ever more precisely will result in little useful information regarding which industries may require assistance to ensure their continued presence in the community.

In a similar vein, Hoover and Giarratani (1984) argue that a region's export activities are not exclusively in the basic (export) sector as it is typically measured.

> It would be more appropriate to identify as basic activities those that are *interregionally footloose* (in the sense of not being tightly oriented to the local market). This definition would admit all activities engaging in any substantial amount of interregional trade, regardless of whether the region we are considering happens to be a net exporter or a net importer. Truly basic

industries would be those for which regional location quotients are either much greater than 1 or much less than 1. (1984, p. 319)

9. The use of censuses is extremely rare. In 1937, *Fortune* magazine conducted an economic base analysis of Oskaloosa, Iowa ("Oskaloosa vs. the United States," *Fortune*, April, pp. 54-62). The study involved a complete census of the city's population to determine the origin and destination of income flows and thus remains one of the most complete applications of the economic base concept to date.

10. This issue raises a simple practical difficulty: should some intermediate inputs, called "indirect primary" inputs by Blumenfeld (1955) and "embodied exports" by Tiebout (1962), be considered basic even though the exporting firm purchased them locally? Exporting firms purchase a variety of inputs, including water, police, fire protection, and labor. Indirect primary inputs could, logically, include the entire nonbasic sector.

Although in this chapter Blumenfeld's contribution is recognized more often than Tiebout's, the latter is equally important. More than any other scholar, Tiebout forged the connections between economic base theory and Keynesian theory. Published in 1962, *The Community Economic Base Study* remains one of the most theoretically complete and practically useful references in the field. In his critique, Tiebout notes weaknesses of the economic base concept: the multiplier is not constant, exports are not the sole source of growth, their importance declines with the region's size, and regions are poorly defined. Tiebout also argues that the comparative cost of a region's production will depend partly on the efficiency of local industries. As a result, the rate of growth is neither strictly exogenous nor solely dependent on export base growth.

11. Hoover also has criticized economic base theory from the perspective of linkages—the connections between industries. Vertical linkages, which exist when one industry uses another's output, can be summarized with the help of the table of interindustry coefficients (Appendix 3.2). Growth can be stimulated by backward linkages—connections that move from demand to supply, back to suppliers of that supply—or by forward linkages—from resources that are turned into products that are sold in the market. These linkages tend to become complementary. A new basic firm, therefore, can attract not only those businesses that supply it, but also those it supplies. Because local services offer important inputs to basic industries, the efficiency of the local sector is critical to export firms. As a result, the primary industry often becomes as dependent on the suppliers, as a group, as the suppliers are dependent on their primary customer.

Such linkages among firms located in close proximity result in cost savings, called *external economies* or, more precisely, external *agglomeration economies*. Hoover argues that the big city is the natural habitat of the small plant because it is most strongly dependent on the services of other plants. These service firms enable small firms to be narrowly specialized and, presumably, more efficient. Small plants concentrate in cities to experience external economies that often depend on the availability of increasingly specialized auxiliary and service firms that, in turn, received their impetus from supplying some main (export) industry. Large metropolitan areas exist, survive, and grow because their business and consumer services enable them to substitute new "export" industries for any that decline or disappear.

12. Five less serious weaknesses beset economic base theory. First, the model ignores specific linkages to other areas by treating all areas outside the region as one undifferentiated rest-of-the-world sector. Second, with price effects absent, the microeconomic forces influencing comparative costs are not addressed. Third, the model does not accord with the reality of regional development in the sense that fast-growing regions would be expected to import rather than export capital (Richardson 1973). Fourth, the theory neglects the role of payments received for reasons other than the performance of work (for example, transfer payments). Finally, while the basic–nonbasic ratio will not necessarily remain stable over time, the model has no means of explaining either the length of the period over which a given external change in demand is expected to influence local economic activity or the internal structural changes that occur as regions grow.

13. Because the theory focuses on one open region rather than on the larger system, the issue of constant-sum or zero-sum growth is ignored. Industrial recruitment efforts in a given community may increase basic employment at the expense of employment in other regions and cities. When plants relocate, there is the very real possibility that one community's gain is simply another's loss. The nation or state as a whole receives no aggregate benefit; hence, the term *zero-sum growth*. In this respect, Blumenfeld's (1955) concern for welfare as the focus of economic development deserves emphasis. The narrow pursuit of jobs puts developers squarely in the business camp concerned about sales. The welfare orientation keeps the focus on the well-being of local consumers.

14. Of course, the utility of surveys depends on their level of accuracy, as well as their ability to obtain all the information necessary to calculate the economic base. The economic base of many regions will consist partially of output consumed within the region by nonresidents; surveys of firms cannot measure this element of the base, since retailers generally do not know the place of residence of their customers. Farness (1989) discusses these problems, along with the reliability of indirect bifurcation techniques when local output is consumed by residents spending externally derived funds.

15. Interested readers are urged to consult Isserman (1980), who provides a more detailed discussion of the theoretical foundation of each technique than we can provide here. We use Isserman's notation.

16. For ease of exposition, we will consider the nation to be the reference area of choice.

17. Note that the use of employment data poses several conceptual problems. Employment data place equal weight on part-time and full-time employment, cannot distinguish between productivity and wage differences across workers in different industries, and do not account for the role of unearned income, including transfer payments, rents, interest payments and profits (Krikelas 1992).

18. Isserman (1980) shows the greatest improvements in estimates of export activity from using disaggregated data come from the shift from division to two-digit-level SIC data. More modest improvements occur as more detailed three- and four-digit data are used.

19. The method was first described by Ullman and Dacey (1960).

20. In Ullman and Dacey's (1960) original discussion of the method, their intended treatment of the regional consumption term was sufficiently ambiguous to

generate what in retrospect appears to have been unwarranted criticism. In particular, Pratt (1968) showed that if the consumption term represented a region's total consumption of good *I*, then every region except the minimum must export while the minimum is assumed to fully satisfy its own internal demand. Therefore, the nation must have net exports. In a subsequent publication, however, Ullman, Dacey, and Brodsky (1969) made clear that the consumption term is an estimate of regional consumption from its own production; there can be regional imports, exports, and cross-hauling. See Isserman (1980) for a more complete discussion of this debate.

21. As noted in section III, embodied exports are intermediate products embodied in direct exports.

22. The five accounts are income and product, input–output, money flow or flow of funds, balance of payments, and wealth. These accounts apply the basic organizing concepts of double-entry bookkeeping and constitute a consistent system of regional economic information. For an excellent explanation of these accounts, see Czamanski (1973).

23. Income and product accounts measure economic flows, usually for a period of one year. They consist of subaccounts for households, firms, governments, capital flows, and rest-of-the-world. They account for all flows that give rise to or use income. The main aggregates derived from these accounts are gross regional product and/or gross domestic product, net regional product, regional income, personal income, and disposable personal income.

24. The import coefficient should include only competitive imports, which are defined as inputs that could have been produced within the region. In national input–output tables, competitive imports are added to domestic inputs to reflect more accurately interindustry flows and production relationships. What are called imports are, in fact, only those commodities that cannot be produced domestically.

25. The stages theory presented by Seers (1963) makes use of these relationships to show why developing areas often suffer considerable economic distress as they attempt to industrialize and engage in international trade.

4

Extensions of Economic Base Theory

Economic base theory remains important for understanding regional economies, in part because of its simplicity. Yet, its simplicity results in its many limitations (as discussed in chapter 3). In this chapter, two economic development theories and three concepts that describe economic relationships are presented as different ways to overcome the four most limiting aspects of economic base theory: its near-term orientation, its exclusive recognition of external demand, its focus on one aggregate multiplier, and its context of one open region interacting with an undifferentiated rest-of-the-world. First, *staple theory* extends economic base logic to address the historical evolution of an undeveloped region over the long term. Second, *sector theory* assumes a mature region that has attained sufficient size to make internal demand more important than external demand as a source of growth. The concepts of *city size*, *linkages*, and *urban hierarchy* represent concepts that can be used to extend economic base theory to account for internal complexity (interindustry relationships) and external complexity (the interregional system). These perspectives set the stage for the richer theories of economic growth and development presented in chapters 5 through 10.

I. Staple Theory

Douglas North (1955) uses export staple theory to examine the historical evolution of regions in order to address development over the long term.

Thus, he extends economic base theory, which best explains near-term growth and decline. His concern is to explain the causes and consequences of secular change, as measured by regional product and regional or per capita income, more than cyclical change in the regional economy.[1] He emphasizes the need to understand a region in terms of its evolving export specialization. Although other sources of growth may exist, the export staple usually is the predominant source. Long-term growth is a function of the quantity and quality of productive factors that are shaped by the export staple and, in turn, produce competitive staple exports.

In writing about economic development in the 1950s, North makes a twofold contribution. He explains and applies staple theory, originally formulated by Harold Innis (1920, 1933, 1940) as a model of the Canadian economy, and cogently criticizes the Hoover–Fisher stages theory.[2] North's application is most appropriate to the economic history of either the United States or Canada, both of which represent undeveloped regions that never experienced feudalism. The point that distinguishes undeveloped regions, such as the United States before 1800, from underdeveloped regions, such as colonial territories, deserves emphasis. Undeveloped regions are places where land and natural resources are available for capitalist development. No established societies exist in the territory at high densities. Underdeveloped regions contained well-established societies before capitalist development was initiated in the region.[3]

Staple theory addresses long-term economic growth and structural change. The staple is an internationally marketable commodity generated by agriculture, forestry, fishing, or mining activities and processed to varying degrees by manufacturing activity. North attempts to extend the staple concept to apply to a region's entire export base as defined in economic base theory. His attempt is only partially successful, however, because the theory works best when applied to export staples, rather than to all exportable goods and services. As Innis observed, the status of the export staple will determine and dominate the development of a region newly integrated into the larger market system.

There are important differences between overlaying capitalist development on feudal European regions and introducing capitalist development as a venture for exploiting an undeveloped region. Capitalists determine what is exportable by gauging comparative advantage and transfer costs. Because demand is exogenous, whereas costs are endogenous, government intervention is focused on reducing transfer costs and driving down production costs in order to expand the market for the export staple. With the development of infrastructure and needed services, firms begin to benefit from external spatial economies. Infrastructure investment, especially in transportation, is needed to improve the competitive position of developing regions. Research and development (R&D) is designed to improve the export staple's technology, thereby reducing costs and deferring diminishing returns. External economies grow up around the export base, specifically through trained labor, complementary industries, and business services, including credit and transport services, that are oriented to the export sector. Capital coming into the region is invested in the known competitive advantage that reinforces the region's specialization; thus, the export staple shapes the entire character of the region's economy.[4]

North describes both the urbanization and the transportation effects of staple-led growth. Unlike the central-place pattern that emerges in densely settled regions where numerous market towns serve local markets, large cities arise in the newly developed region. For example, most major U.S. port cities have grown as physical and commercial interruptions or breaks in transportation. Extensive transactions occur in these locations, spurring the growth of financial and legal services. Public investments are geared to reducing transfer and transport costs so as to increase the size of the market for the staple.

Growth of the export staple usually leads to related manufacturing growth. Considerable new industry can result from the successful export staple. First, materials-oriented industries (such as sugar refining) result in vertical integration (sugar cane or beets, for example, are first processed crudely into sugar and molasses, then into more refined sugar products). Second, nonbasic industries serve the growing local market. Third, industries arise to supply the staple-related manufacturing and primary sectors. Finally, footloose industries, although not directly tied to the export staple, are attracted to the region because of the growing size of its market. They serve as an outlet for locally accumulated capital. Thus, industrialization occurs naturally; government intervention is not necessary to spur local private investment. In the long term, growth gradually reduces differences among regions that specialize in different staples. With maturation comes greater diversity among local economies and more similar local economic structures.

Application

Staple theory provides a general strategy of development by recognizing the connections of the economic base to the political superstructure. Economic developers should continue to build on and improve the export staple as long as it remains competitive in the larger economic system. Strengthening the existing specialization may be a more sensible strategy than attempting to diversify the economic base. The idea is to do what one knows best.

Eventually, footloose economic activities will be attracted to the area if its market achieves sufficient size or if it offers urbanization economies that can be exploited by other exporters.

Government intervention is rationalized as a means of improving the competitiveness of the export staple. Public investments in R&D, infrastructure, training and so on are justified as long as they lead to staple production at lower cost. Concern about greater diversity should be deferred until a strong industrial complex and the related infrastructure are created to support the staple export. Economic developers, then, should not pursue job creation through business development strategies. Instead, they should fulfill the traditional roles of government, which are intended to support capitalist development and thus sustain the export-staple industrial complex. In other words, to increase the efficiency of local firms, economic developers should provide public infrastructure, education and training, new scientific information, and other public goods.

Columbus, Ohio, for example, is not a local economy whose growth was dominated by an export staple. Ever since the founding of the city, state government has remained a central economic specialization. Entrepreneurs won the competition for the state capital by proposing an ambitious real estate development project in an undeveloped area. The government sector, however, did more to build up demand in the local market than to stimulate linkages to the external economic system. The theory is more helpful in understanding the economic history of many port cities and other major cities in North America, which grew initially by providing the physical and commercial transitions between water- and land-based transportation that connected an expanding inland economy to foreign markets.

Although staple theory has limited application, it is important and compelling in one respect: it orients economic developers to the economic history of their local economy. Developers should study and comprehend local economic history, for two basic reasons. First, knowledge of economic history provides the grounding needed to apply appropriately and skillfully more abstract, ahistorical models of economic growth. Second, and more important, this understanding will help developers work more successfully within the constraints imposed by local values, politics, and wealth.

II. Sector Theory

Sector theory argues that, over time, the relative share of production in each major sector will change in the region. Thus, because it is relatively simple and testable, the theory presents an attractive way to understand local economic development. In sector theory, the economy is divided into three highly aggregated sectors: primary (agriculture, forestry, fisheries), secondary (manufacturing and mining), and tertiary (trade and services). Due to the income elasticity of demand for primary, secondary, and tertiary products, and to the fact that technological change leading to productivity improvements is more pronounced in primary and secondary sectors, the region becomes specialized in primary, then secondary, then tertiary products.

In contrast to economic base theory and staple theory, which emphasize external economic relationships, sector theory focuses on the internal structure of the economy. According to Perloff, "sector theory focuses on internal rather than external development; economic growth is seen as primarily an internal evolution of specialization and division of labor, although external shifts in demand are not ruled out as of no importance" (Perloff et al. 1960, p. 59). Internal development through specialization and division of labor paves the way for favorable external trading relationships. The structural relationships among the three sectors evolve as income per capita increases in the economy over time. Development is identified with internal changes in the economy that lead to increased specialization. Conversely, the logic of outwardly focused economic base or staple theory argues that funds earned from trade are fundamental and necessary to bring about internal development.

Sector theory is attributed to Clark and Fisher, whose writings were published between the mid-1930s and the early 1940s. In *The Conditions of Economic Progress*, Clark (1940) observes that a high level of real income per capita is associated with a high proportion of the labor force in tertiary industries. Examining national economic structure, Fisher (1933) makes the same observation. Sector theory was developed to explain this empirical phenomenon. The economy is supposed to shift from "lower" order to "higher" order economic activities as economic growth proceeds. Based on their empirical research, Clark and Fisher expect shifts to occur from primary sectors through secondary sectors and on to tertiary sectors.

The income elasticity of demand for the products of different sectors drives the sectoral shifts in production. Increases in labor productivity support the changing sectoral allocation of the labor force. As income *per capita* increases, the demand for manufactured goods will exceed the demand for agricultural and other primary products. Subsequently, the demand for services predominates, and the service sector becomes the largest regional sector.

At the same time, growth-creating investment introduces better technology, which allows factor substitution. Generally, technological progress will be labor-saving in the primary and secondary sectors. Primary sectors and mining industries are expected to experience diminishing returns, while manufacturing should experience constant or increasing returns to scale. Production relationships encourage the expansion of manufacturers relative to producers of primary products. Producers continue to introduce labor-saving

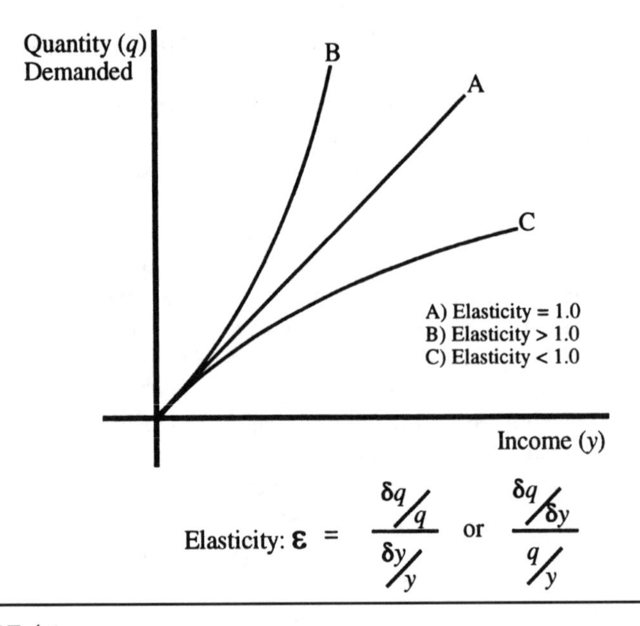

$$\text{Elasticity: } \varepsilon = \frac{\delta q / q}{\delta y / y} \quad \text{or} \quad \frac{\delta q / \delta y}{q / y}$$

FIGURE 4.1
Income Elasticity of Demand: Constant, Increasing, and Decreasing

technology in order to respond to shifts in demand. These productivity improvements make labor available to the tertiary sector, since they are more likely to displace labor from primary and secondary sectors. As a result, the share of total employment is lowest in primary sectors and highest in tertiary sectors.

Income elasticity of demand (compared to price elasticity which relates change in quantity demanded to change in price) relates the demand for a commodity to income, as shown in figure 4.1. Except for inferior goods, the quantity demanded should increase with aggregate income. If the rate at which demand increases with income is constant, income elasticity equals 1.0. Increasing and decreasing rates of change in demand refer to elasticities of greater than or less than 1.0, respectively. Figure 4.2 shows how income elasticity may affect the demand for commodities of each sector. Starting at the point of origin, the demand for agricultural goods peaks quickly, then levels off; manufacturing goods increase in level but at a declining rate; services show an increasing level at a constant or increasing rate.

Although sector theory was sharply criticized in the 1950s by economists who were using developmental models that stressed external growth, it remains a respectable explanation of economic progress, for several reasons.[5]

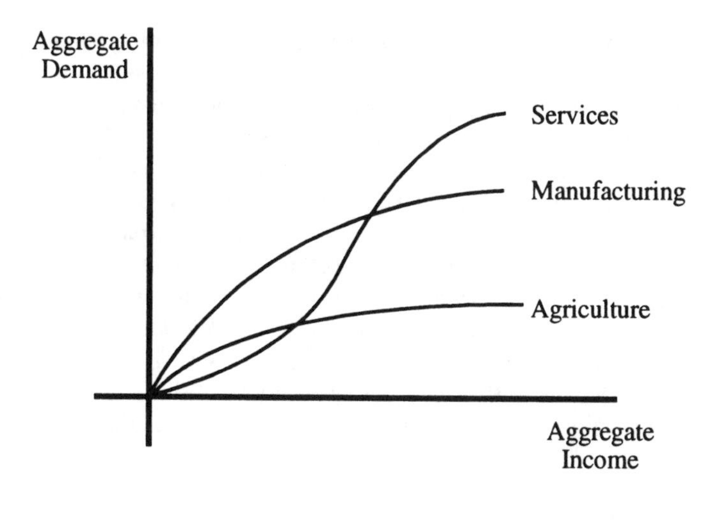

FIGURE 4.2
The Effect of Income Elasticity on Commodity Demand

First, it analyzes economic structure and structural change, not simply the income or production flows of an existing economic structure. Second, although it is relatively easy to understand, the theory has rich applications, as described below. Third, the theory deals with both demand and supply.

On the demand side, the sectors can be placed in a hierarchy in order to consider the evolution of an economy. Food and shelter are the most pressing needs. The primary sector initially is dominant because its products are necessities. Demand for this bundle of goods is income inelastic. As an economy becomes more efficient in the production of the goods needed to sustain life, surplus labor becomes available for other pursuits and is put to work in the secondary sector, manufacturing goods. Demand for these goods is more elastic than it is for primary goods. With innovation and productivity increases in the secondary sector, surplus labor again becomes available and moves to provide services for those employed in the primary and secondary sectors. Demand for services is highly elastic. As an economy evolves, employment increasingly is provided by the higher-order sectors. Thus, the service sector can enhance the range of economic opportunities within a community. A large service sector can support more diverse economic activities. In many metropolitan areas, increasing service sector size often reflects greater ability to attract, retain, and improve manufacturing.

On the supply side, both Clark and Fisher see productivity gains as the direct result of increased specialization. These gains are both qualitative and quantitative. With increased specialization come increased expertise, increased rates of innovation, and more creative uses of labor and capital. Internal structural development that increases with greater specialization increases the number of activities the economy can support simultaneously. The diversity of economic opportunities gives all members of society greater choice to pursue the work for which they are most capable. The internal development of economic structure is a mark of the economy's growing sophistication (Fisher 1933). Sector theory, then, may be used for predictive and prescriptive purposes, if one is willing to assume that less-developed economies will follow the path of more-developed economies. The size of the tertiary sector is a crude indicator of the level of development, and the growth rate of the tertiary sector may be a dynamic indicator of development.[6]

Application

Economic developers tend to consider export sector expansion and import substitution as the best strategies to encourage economic growth. Sector theory reminds us that furthering the internal evolution of the local economy is an equally sound path to economic development. Internally focused development strategies can lead to progress over the long term; external development strategies may promise more rapid economic growth but also higher risk of failure and instability.

Nevertheless, sector theory is too crude to be used as a strategy for encouraging economic development in its original three-sector form; sectors must be examined at a finer level of detail to gain useful insights. The tertiary sector, in particular, has become too diverse to be treated as a single category. Since the theory was introduced, whole new service industries have been created, and services are now consumed on a mass scale. New service industries range from cellular telephone service to copying centers to pizza delivery shops. The motivating forces behind service sector growth are as diverse as the new services themselves, ranging from outsourcing, increased labor force participation rates, and more recreational opportunities. Because the spectrum of today's service sector is so broad, its specific role in shaping the internal structure of an economy is not easy to define.

To use sector theory more effectively, the service sector may be divided into five broad categories: (1) distribution, (2) trade, (3) business services, (4) education and health services, and (5) other public, nonprofit, and consumer services.

Distribution and trade are not identified as advanced services (see Appendix 4.1), notwithstanding considerable change within these sectors. Business services are important, especially for newer businesses, because they can reduce overhead costs and allow greater flexibility. These services have grown rapidly, partly as a result of outsourcing from primary and secondary sectors, including word processing, advertising, legal services, and accounting. Regional economies in which business services are overrepresented may be functioning as advanced service centers. Education, health, and governmental services can spur regional growth by serving both local and external markets. They shape the essential character of university cities, medical centers, and state capitals. On the other hand, consumer services usually experience modest growth and are not very important unless the local population is extremely affluent.

Developers may extend this detailed analysis to all sectors in order to anticipate growth and decline of local industries and plan for ways to respond to expected change. Comprehensive scope requires detailed structural analysis of the metropolitan economy. Input-output analysis, introduced and described in chapter 3, is one vehicle for conducting structural analysis.[7] To conduct this structural analysis, the model should be generalized to examine (1) the income elasticity of demand for goods at different levels of aggregate income or per capita income, and (2) the expected technology by sector. For U.S. metropolitan areas with different industry mixes, aggregate income growth will have differential effects on local production, since income elasticity stimulates some sectors more than others. Productivity gains also will depend on industry mix. The local labor freed for new activities may find new employment locally, or it may migrate to other areas.

Alternatively, local economic developers may prefer to focus detailed analysis on the productivity trends of a few major industries and the income elasticity of major products. They may find this analysis useful in attending to existing businesses and industries. Such efforts typically include annual calls or visits to important employers, assistance in dealing with government agencies, advocacy for public improvements that benefit existing companies, business appreciation days and awards, and many other services supportive of existing businesses. These activities usually are justified on recruitment grounds; the existing industries in one locality are the recruitment targets of other areas. Thus, it makes sense to care for local companies for fear of losing them.

Sectoral analysis offers powerful insights into the relative prospects of existing industries. These insights would help local economic developers ration scarce resources in order to assist the most strategically important sectors. Currently, however, such applications are uncommon; the productivity of local industries and the well-being of local consumers take a back seat in economic development practice to the expansion of exports or the reduction of imports.

Industrial targeting is a popular aspect of the recruitment strategy. Targeted industries are attractive typically because they pay relatively high wages, have minimal negative impacts on the environment, and are forecast to grow nationally. Many industries are growing because they sell income-elastic products. Such determinations of growth prospects are made at the national level, for example, by economists in the U.S. Department of Commerce who produce the publication, *Industrial Outlook*, or provide other forms of economic intelligence. Local economic developers can use this information to anticipate and facilitate structural change.

In conclusion, economic developers often limit their theoretical insights about the local economy to those provided by economic base theory. This chapter is designed to demonstrate how much ground developers can gain by mastering four extensions of that theory. In essence, these insights can give developers much better understanding of their local economy. Staple theory requires paying attention to the historical evolution of local economic structure. Sector theory emphasizes internal economic relationships and future development paths. As discussed in Appendix 4.1, structural relationships may be analyzed more fully, using input-output models, while the concepts of diversity and centrality can help developers comprehend the significance of city size and the city's location in the urban hierarchy.

Discussion Questions

1. Why would popular strategies to increase investment and employment—attraction, creation, retention, and expansion (ACRE)—not find support from the perspective of staple theory?
2. In general, which economic development strategies are recommended from the staple theory perspective?
3. Is/are the industrial complex(es) in your area the outcome of staple exports over many years?
4. List important insights that can be gained about your area from an understanding of its economic history.
5. Is the service sector diverse and sophisticated in your area? Will its development lead to increases in local economic growth and development?
6. If the United States is now dominated by service industries, will services continue to grow relative to other sectors? (In the international context, one could argue for an increasing share for agriculture, a constant share for manufacturing, and a declining share for services.)
7. Although aggregate income continues to increase, real household income has declined in some areas since the 1980s. Does sector

theory require per capita income increases, or are aggregate increases sufficient? Does the model work in reverse if aggregate or per capita incomes decline?

8. Will productivity improvements in information flow and management displace employment in tertiary sectors? Will new labor-intensive industries arise?
9. Should income (demand) change be gauged at the national or the global level? What is the proper level of analysis?
10. Is your area dominated by one functional specialization or by many? How do you define and measure functional specialization?
11. Are important industries relatively independent, or are they part of one or two industrial complexes? What evidence would you provide to demonstrate this situation?
12. How should economic developers treat their metropolitan area's functional specialization? Is it a given, such that the basic strategy is to compete successfully with comparable areas? Is the basic strategy to change/improve the existing functional specialization? If so, how?
13. Which economic advantages could accrue to a city due to its population size?
14. List all the factors you believe give rise to agglomeration economies in your area. List the factors that generate diseconomies of agglomeration.
15. Is your area in Noyelle's typology shown in table 4.1? Do you agree or disagree with its classification? Why?
16. Is achieving greater economic diversity an important economic development objective in your area? How can you pursue that goal, given the existing functional specializations and industry mix?

Appendix 4.1

CITY SIZE, LINKAGES, AND URBAN HIERARCHY

City Size

One of the most deceptive concepts in regional theory is city size. The size of the local labor market or metropolitan area has been implicitly considered in presenting economic base theory, staple theory, and sector theory. Economic base theory is more applicable in the context of a small community with a single dominant specialization; this city probably must export or die. Staple theory takes a historical perspective to explain how great cities can arise in formerly undeveloped territories when the staple product remains competitive in external markets over time. Sector theory was conceived in the context of mature national economies and so presumes an economy of major cities that have attained substantial size. What, therefore, does city size actually signify?

The city consists of people living and working in one place. Local labor markets and metropolitan areas in the United States, which we use as the spatial units to explain theories of economic growth and development, vary widely according to geographic coverage and density. Yet the unifying, relevant feature here is the existence of the metropolitan community—a group of contiguous counties that create a single "commuter shed" that contains economic interactions between and among households, firms, and government entities.

What are the important differences between cities of different size? First, large cities provide large internal markets. Large size, in itself, attracts producers because the resident households and firms combined have more aggregate income to spend on goods and services.[8] Therefore, a positive feedback mechanism connects larger size to growing demand, which attracts more firms and increases size further. Although such growth is not without constraints, larger cities clearly generate sufficient internal demand to stimulate economic activity. The basic assumption of sector theory and the context assumed by Blumenfeld (1955) in his criticisms of economic base theory are appropriate for major metropolitan areas but not for smaller communities.

Second, large cities have sizable labor forces, which encourage substantial division of labor and greater productivity. At the other extreme, company towns have all workers in their relatively small labor force directly or indirectly supporting the leading firm. Over a considerable population size range, the city is able to support only one specialization efficiently. The actual size attained by single-specialization cities varies considerably and depends largely on the internal economies of scale, size, or scope enjoyed by firms in the core sector. Yet large size is clearly associated with greater specialization; a city with one million people can support more specializations than a city of 100,000.

The *economic diversity* of a city can be defined in reference to its specializations. As additional relatively independent specializations co-locate, the area becomes more diverse. Economic diversity is the presence of multiple specializations. This definitional point deserves emphasis because the diversity literature is so confusing. In some instances, diversity is viewed as the absence of specialization. This view is typical of studies that use the coefficient of specialization or similar specialization measures to indicate degree of specialization/diversity. In other instances, modern portfolio theory (MPT) is used in regional analysis. With MPT, the growth and temporal stability of regional industries become the focus of analysis, and the degree of correlation between industry-specific growth rates determines regional "diversification" possibilities.[9]

The proper conceptualization of diversity recognizes the possibility of adding specializations as the labor market or city size grows. Detailed analysis of industrial structure is needed to determine the core industries and related sectors that support one or more industrial complex(es) in one place. Those metropolitan areas with the greatest number of industrial complexes are the most diverse. Population size is a crude indicator of economic diversity because larger size is a necessary, but insufficient, condition for greater diversity. For example, the following metropolitan area pairings have roughly similar 1995 population size but considerable differences in economic diversity measured as the number of two-digit sectors with employment and uniformity of the employment distribution across two-digit sectors. The first area listed is more diverse than the second: Philadelphia–Detroit, Atlanta–Houston, Baltimore–St. Louis, and Denver–San Jose.

The other structural feature associated with city size is *centrality,* a concept that refers to the economic importance of a place in relation to other areas. Centrality is applied to describe the urban hierarchy given below. Larger cities are more competitive than smaller ones to the extent that larger size is the result of greater economic diversity and centrality. The only advantage of size is the concentration of population that results in larger markets.

Linkages

Input-output analysis is a useful technique that can reveal interindustry linkages and their influence on regional growth and change. As described in chapter 3, it is both an accounting system and an analytical model. It can be used to extend economic base theory, as shown in Appendix 3.2. Compared to the economic base multiplier, the industry-specific multipliers derived from the Leontief inverse matrix give a much more detailed picture of the ways external stimuli affect the local economy. These industry-specific multipliers summarize the extent of linkage within the regional economy. Local multipliers increase as interdependence increases over time and as competitive imports decline.

The input-output model can be used more generally to analyze the structure of the regional economy, both over time, as noted in the discussion of sector theory in section II, and in comparison to other regions. For one region in any year, regional industries can be ordered from those most exclusively providing output to final demand, to those industries that produce entirely for other local industries. This ordering, called triangulation, shows the extent of industrial hierarchy (versus circularity) in the regional economy. The more typical pattern for major metropolitan areas is high interdependence among several groupings of industries, with few linkages between the groupings. Structural analysis of national input–output tables has led to the identification of consumer goods clusters, capital goods clusters, and universal intermediaries.

Edgar Hoover (1971) presents a well-known formulation that combines interindustry relations and locational factors to suggest how linkages can stimulate regional growth. The industry location effect of horizontal linkages between regional enterprises is one of mutual repulsion (a limit on growth), whereas vertical linkages are characterized by mutual attraction (a stimulator of growth). Alternatively, the relationship between local enterprises has both a supply-constraint dimension (horizontal linkage) and a demand-creation dimension (vertical linkage). Hoover identifies two types of vertical linkage: backward and forward.

Horizontal linkages between economic activities in a region reflect the fact that businesses, nonprofit enterprises, and governments compete for scarce sites, physical resources (water and clean air), and labor. The existing industries that fully utilize currently available local resources preclude the expansion of other industries in the near term (due to supply constraints). Demand for resources drives up costs, which in turn creates an incentive for businesses and other activities to spread over the landscape rather than concentrate in specific cities and regions. The notion of mutual repulsion affects the locational tendencies of business enterprises. For example, convenience retailing enterprises fight for local market share.

Vertical linkages characterize the relationship between firms in the same product or input–output chain. Suppliers and buyers of the same products benefit from reduced transfer costs that result when firms in the same product chain concentrate in one place, thereby offsetting the effects of horizontal linkage. Each industry (column) of a regional input–output model portrays the degree of connectivity in the local economy. Over time, growth among buyers can stimulate the attraction or expansion of suppliers; such growth is due to what Hoover calls "backward linkages." Conversely, expansion among suppliers can facilitate the attraction or expansion of buyers—growth due to forward linkages.

Regional analysts have devised several techniques for examining vertical linkages to identify groups of related industries, or industrial clusters (Czamanski and Ablas 1979). In addition to triangulation, these include graph theory, principal components/factor analysis, and statistical cluster analysis. Essentially, these approaches involve using the technical coefficients from national or regional input–output tables to identify measures of the strength of linkage between sectors. Patterns in the linkages are then observed in order to identify groups of technologically related industries or extended product chains; typically, the groups constitute one or more major final market producers together with their first-, second-, and third-tier supplier sectors. The techniques are most effective when applied to manufacturing industries where material flows are most significant.

Researchers have studied whether sectors related through formal production linkages have a tendency to co-locate in space. In the early literature, clusters that exhibited a high degree of geographic concentration were defined as industrial complexes (Czamanski 1974, 1976). In the recent literature, the concept of the complex has become confused with that of the cluster; while, at the same time, the meaning of the latter has been broadened significantly. For example, Michael Porter (1990) defines industrial clusters as strategically related industries that have a tendency to concentrate in particular regions. Porter's work is suggestive of the role of local linkages in enhancing national competitiveness and international trade. However, his concept (and, subsequently, that of other researchers) of an important, interindustry relationship extends well beyond trade in physical goods, to include shared business-related institutions, informal cooperation, and shared human resources (Rosenfeld 1995). On the basis of this work, many state and local economic developers have tried to identify and develop their own strategic clusters (Doeringer and Terkla 1995). Unfortunately, although this focus on clusters has become extremely popular in economic development practice, methods of analysis remain rudimentary, often involving little more than the use of location quotients to indicate regional specializations (Feser and Bergman 1998).[10]

The interfirm relationships that underlie the broader meaning of a local industrial cluster are consistent with the general concept of spatial external economies. External economies, which may be either positive or negative (if the latter, they are called diseconomies), refer to side effects of one producer's activity on another's. Early analysis of such economies distinguished between technological and pecuniary external economies (Viner 1931, Scitovsky 1954). The former are effects that are transmitted outside the price mechanism and thus lead to deviation between private and social marginal cost (market failure). The latter refer to general price effects—for example, when one industry's output lowers the cost of inputs for other industries—that indicate how industries' fortunes may be related yet remain fully reflected in the price mechanism. Pecuniary external economies do not generate market failure and are not the subject of significant analysis in mainstream economics.

In the urban and regional development literature, external economies are referred to as agglomeration economies. These are reductions in costs or improvements in productivity for individual firms, which come about through the concentration of industry and other activities in particular locations. Building on Weber (1929), Hoover (1937) distinguished between two specific types of agglomeration economies: (1) localization economies, or cost reductions for firms in a given industry that resulted from the spatial concentration of that particular sector; and (2) urbanization economies, or cost reductions for all firms in a given location, which come about as economic activity in general expands in that place. Until recently, the literature on agglomeration economies focused largely on the benefits of urban and industry size rather than with the specific synergies firms enjoy via spatial juxtaposition (Feser 1997). These synergies were generally a secondary consideration; most of the time, they were discussed as justification for the hypothesis that firms were more productive in larger cities and in industrial complexes.

More recently, research has shifted back to a focus on the sources of economies or mechanisms by which firms influence each other's competitiveness. Many of these studies build on Marshall's analysis (1961 [1890]) of external economies that arise in what Marshall called industrial districts. Marshall identified the three key benefits of proximity between firms: (1) greater availability of inputs, (2) presence of deep labor pools, and (3) greater exchange of information and knowledge of new techniques and innovations. Marshall emphasized (although he did not fully explore) the important roles of industrial structure and institutions, in addition to the size of the place or industry, as influences on the degree of beneficial spatial effects. Modern studies of industrial districts are discussed in greater detail in Chapter 10.

Urban Hierarchy and Central Place Theory

The final extension of economic base theory introduces spatial complexity. Instead of a theory that considers one subject region interacting with the rest-of-the-world (ROW), the subject region can be viewed as one region interacting with many other regions. For simplicity, this interregional system can be reduced to a closed system of two regions, as in interregional trade theory (chapter 7); it can also be treated as a multiregion system.

For many years, beginning in the 1940s, regularities in the city-size distribution of nation-states were recognized. The "rank-size rule" popularized among geographers refers to the clear, positive association between the rank-order of an urban area and its size; the relationship between the logarithm of city size and the rank-order of cities was approximately linear. These distributions were later described as log-normal city-size distributions in contrast to primate city-size distributions. For example, more developed countries such as the United States had a log-normal distribution—the representation of cities along the entire size range, with an increasing number of places as city size decreased. On the other hand, many less-developed countries had a primate distribution—one large capital or port city, virtually no intermediate-sized cities, and many small cities. These ad hoc observations led to empirical tests of the relationship between level of economic development and distribution of city size. No clear relationship was found.[11]

In the theoretical realm, it has been shown that central place theory can be the basis for an urban hierarchy that follows a log-normal distribution.[12] Central place theory models the way in which consumer goods can be distributed most efficiently within the space economy. The key concept is the *spatial range* of any good or service. Consumer products for which there is frequent and consistent demand can be provided everywhere in the "lowest order" places. As demand declines, the range of the commodity increases because an ever-increasing number of consumers is needed to support "higher order" goods and services. At the top of the urban hierarchy is the highest-order central place, which provides all consumer goods and services. As one moves to lower- and lower-order places, fewer and fewer consumer goods and services become available locally.

In its original formulation, central place theory was best used to explain the distribution of market towns in largely agricultural regions. Palm (1981) presents a synthesis that begins with central place theory. The central place hierarchy works best for relatively closed systems where centers serve the larger territory and trade flows within the system. Long-distance trade among more open systems leads to major settlements that support merchant capitalism. These places are the physical and commercial breaks in transportation referenced in the presentation of staple theory. Neither pattern describes the current U.S. urban system well. Palm summarizes the work of Pred (1976), who suggests a much more complex pattern based primarily on corporate control. The intraorganizational linkages of multilocational firms are more important than the spatial linkages among proximate economic activities. The former can be viewed as an overlay that reflects control functions more than distribution activities.

Noyelle and Stanback (1984) have updated the central place concept, consistent with Pred, to apply to the provision of higher-order services in the contemporary U.S. economy. The authors bring together the concept of centrality with the concept of economic diversity described above. They argue that in the modern interregional system—the global economy—the places of highest order are centers of command and control, places some observers would associate with world-class cities. Command and control centers are places where important strategic decisions are made, which direct the most powerful financial and nonfinancial corporations, government agencies, and nonprofit organizations. Centers of production, distribution, and consump-

TABLE 4.1

140 Largest SMSAs Classified by Type and Size, 1976

Diversified advanced service centers						*Specialized advanced service centers*					
NATIONAL			REGIONAL			FUNCTIONAL NODAL			GOVERNMENT AND EDUCATION		
1[a]	New York	1[b]	4	Philadelphia	1	5	Detroit	1	8	Washington, DC	1
2	Los Angeles	1	6	Boston	1	13	Pittsburgh	1	39	Sacramento	3
3	Chicago	1	10	Dallas	1	16	Newark	1	48	Albany	3
7	San Francisco	1	11	Houston	1	24	Milwaukee	2	54	New Haven	3
			12	St. Louis	1	31	San Jose	2	64	Springfield, MA	3
SUBREGIONAL			14	Baltimore	1	36	Hartford	2	77	Raleigh-Durham	4
41	Memphis	3	15	Minneapolis	1	38	Rochester	3	81	Fresno	4
45	Salt Lake City	3	17	Cleveland	1	40	Louisville	3	82	Austin	4
46	Birmingham	3	18	Atlanta	2	44	Dayton	3	84	Lansing	4
52	Nashville	3	21	Miami	2	47	Bridgeport	3	85	Oxnard-Ventura	4
53	Oklahoma City	3	22	Denver	2	50	Toledo	3	88	Harrisburg	4
56	Jacksonville	3	23	Seattle	2	51	Greensboro	3	89	Baton Rouge	4
58	Syracuse	3	26	Cincinnati	2	57	Akron	3	90	Tacoma	4
65	Richmond	3	28	Kansas City	2	62	Allentown	3	99	Columbia, SC	4
69	Omaha	3	30	Phoenix	2	63	Tulsa	3	111	Utica	4
91	Mobile	4	32	Indianapolis	2	67	New Brunswick	3	112	Trenton	4
101	Little Rock	4	33	New Orleans	2	70	Jersey City	3	113	Madison	4
106	Shreveport	4	34	Portland	2	75	Wilmington	3	117	Stockton	4
110	Des Moines	4	35	Columbus	2	78	Paterson	4	130	South Bend	4
114	Spokane	4				86	Knoxville	4	140	Ann Arbor	4
120	Jackson, MS	4				96	Wichita	4			
						100	Fort Wayne	4			
						103	Peoria	4			
						137	Kalamazoo	4			

Production centers									*Consumer-oriented centers*		
MANUFACTURING			MANUFACTURING (CONTINUED)			INDUSTRIAL–MILITARY			RESIDENTIAL		
27	Buffalo	2	116	Reading	4	20	San Diego	2	9	Nassau	1
42	Providence	3	119	Huntington	4	37	San Antonio	3	19	Anaheim	2
59	Worcester	3	124	Evansville	4	49	Norfolk	3	76	Long Branch	3
60	Gary	3	125	Appleton	4	87	El Paso	4			
61	NE Pennsyl.	3	131	Erie	4	97	Charleston, SC	4	RESORT-RETIREMENT		
71	Grand Rapids	3	134	Rockford	4	102	Newport News	4	25	Tampa	2
72	Youngstown	3	136	Lorain	4	121	Lexington	4	29	Riverside	2
73	Greenville	3				123	Huntsville	4	43	Ft. Lauderdale	3
74	Flint	3	MINING-INDUSTRIAL			126	Augusta	4	55	Honolulu	3
80	New Bedford	4	83	Tucson	4	127	Vallejo	4	68	Orlando	3
92	Canton	4	105	Bakersfield	4	128	Colorado Springs	4	79	West Palm Beach	4
93	Johnson City	4	118	Corpus Christi	4	132	Pensacola	4	95	Albuquerque	4
94	Chattanooga	4	129	Lakeland	4	133	Salinas	4	108	Las Vegas	4
98	Davenport	4	135	Johnstown, PA	4				122	Santa Barbara	4
104	Beaumont	4	138	Duluth	4						
107	York	4	139	Charleston, WV	4						
109	Lancaster	4									
115	Binghamton	4									

Source: T. Noyelle, 1983. The rise of advanced services (Table 3). *Journal of the American Planning Association* 49, pp. 280–290. Reprinted with permission from the *Journal of the American Planning Association*.

a. 1976 population rank.

b. Population size groups: Size 1: over 2 million population; size 2: between 1 and 2 million; size 3: between 0.5 and 1 million; size 4: between 0.25 and 0.50 million.

Data Sources: Noyelle and Stanback, 1983; and U.S. Bureau of the Census, *Current Population Report*, Series P-25.

tion follow in decreasing order of importance. This ordering follows from the growing importance of marketing, finance, product development, personnel management, and government/public relations in relation to production and distribution activities. The interaction of centrality and diversity generates the typology of metropolitan areas shown in table 4.1.

Because infrastructure usually complements an area's most prevalent economic activities, it should vary according to its place in the urban hierarchy. Command and control centers need good air travel facilities and advanced telecommunications for effective corporate planning and management. These centers attract major airport hubs and communications infrastructure. Production and distribution centers require good access to the interstate highway system, water ports, railroads, and, often, air cargo facilities.[13]

Notes

1. Tiebout (1956a, 1956b) criticizes North's argument. His criticisms reiterate the general criticisms of economic base theory presented in chapter 3, rather than directly address export staple theory. Therefore, Tiebout misses North's essential point, which is to explain growth as a long-term historical process. Tiebout is concerned primarily with near-term cyclical stability.

2. Hoover and Fisher (1949) present an idealized version of capitalist development based on the economic history of Western Europe. In their theory, sectors and stages interact. This growth model describes an economy moving through specializations, from agriculture to manufacturing to services. Industrialization overcomes diminishing returns in agriculture and propels the economy forward.

Hoover and Fisher (1949) segment the development process into five distinct, sequential stages: (1) the self-sufficient subsistence economy based on agriculture, (2) local specialization based on trade and facilitated by transport improvements, (3) cash-crop farming, mining and manufacturing to exploit opportunities in interregional trade, (4) "forced" industrialization as regional population increases and agriculture reaches diminishing returns, and (5) development of exporting tertiary sectors, which involves outflows of capital and skilled personnel. As population increases and transportation and communications improve, division of labor based on comparative advantage leads to increasing specialization and trade. One unique feature of this model is its attention to the space economy. Hoover and Fisher are regional economists who think through the spatial structure implications of various economic stages.

North (1955) begins his article with a broadly based critique of the Hoover-Fisher model of development that he portrays as a descriptive theory that lacks explicit causal mechanisms. First, the Hoover-Fisher theory is not general; it applies only to parts of Western Europe. Undeveloped regions with low population and no feudal institutions are fundamentally different from populated, feudal economies. In the United States and Canada, long-distance trade in world markets was the sustained stimulus for growth, much more than opportunities for local trade. Second, regions need not industrialize in order to progress; progress depends on the export staple and on capital flows that seek their highest reward. Hoover and Fisher incorrectly suggest the need to force industrialization or to balance economic growth.

3. Neoclassical economists argue that underdevelopment results from the imperfect and partial introduction of capitalism in such areas, whereas neo-Marxists argue that underdevelopment is the direct result of capitalist development.

4. North notes that high-income elasticity of demand for the staple can lead to rather violent cyclical fluctuations. In contrast, sector theory, to maximize growth, advocates pursuing sectors with high-income elasticity. North stresses the risk side, while sector theory emphasizes the reward.

5. On the negative side, one could make four points. First, both Clark's (1940) and Fisher's (1933) explanation is based on the apparent relationship between service sector size and the wealth of more-developed countries. This relationship may be one effect of development rather than its cause. Second, the division of economic activity into three highly aggregated sectors begs the question of what is going on within these sectors. For example, Bauer (1951), referring to less-developed economies, questions the hierarchical organization of the three sectors. He notes that many impoverished nations have a large percentage of the labor force involved in service sector distribution activities. This employment is largely a result of the failure of secondary sector growth. Service sector growth may occur when women enter the labor force at higher participation rates. Services formerly provided within households—for example, child care, cooking, and cleaning—may now have to be provided in the informal sector. More paid employment may not result in higher levels of well-being. Third, sectoral classification is difficult. Classifications based on returns to scale may be reasonable, but income elasticities for any sector are not constant over time or space. Elasticities can vary from region to region and from year to year. Finally, sector theory takes the nation-state as its unit of analysis. The theory must be modified to explain the evolution of metropolitan areas. It becomes more applicable as the area becomes more self-sufficient and closed.

6. Miernyk's (1977) empirical study suggests the limited utility of the three-sector model. He tests the relationship between economic structure and the growth of per capita income with data for southern states in 1940 and 1975; the relationship is weaker in 1975, compared to 1940. He finds no correlation between relative income level and manufacturing employment.

Miernyk's study is limited, for two reasons. The theory refers to longitudinal changes in nodal regions; Miernyk looks at states in cross-section. He uses employment data, which measure the quantity of one input, rather than the value of sectoral output. Productivity gains within sectors may increase the aggregate value of labor, although the number employed may decrease or remain the same. Employment growth may reflect the substitution of cheap labor for another input, rather than aggregate growth.

7. Structural input–output analysis can suggest ways to encourage interindustry linkages based on technology or spatial proximity. This analysis provides support for the strategy of building industrial complexes. Input–output has also been used by large corporations and national governments as a powerful economic planning tool.

8. Some theories make use of the empirical finding that large cities also have higher per capita incomes and higher costs of living (for example, see the product-cycle theory in chapter 8). City size has also been used as a surrogate for more or better information, more technical skills, and more entrepreneurial talent, as well as larger markets.

9. Michael Conroy (1975) first used MPT to analyze employment growth and fluctuations in employment over time for U.S. metropolitan areas. From his and subsequent work, an area's growth and stability are outcomes related, respectively, to reward and risk. Diversity and diversification are often defined in these terms. See Malizia and Feser (1994) for a summary of this literature and further discussion.

10. See Tremblay (1993), Anderson (1994), and Morfessis (1994) for applications of industrial cluster strategies and techniques.

11. No relationship was found in this research in part because nation-states were used to analyze city-size distributions. More log-normal relationships would have been found if true systems of cities had been analyzed. For example, all Commonwealth countries and Great Britain have a combined city-size distribution that is much more log-normal than the distributions of any constituent country. The importance of system "closure" is demonstrated in Vapnarsky (1967).

12. The rank-size distribution can be generated by Lösch's (1954) system of nested hexagonal markets, as well as central place theory. The central place theory, rank-size regularity, is addressed in Beckman (1958), Parr (1970), and Beguin (1985).

13. Suarez-Villa (1988) integrates concepts from sector theory and the urban hierarchy. He also: addresses the spatial, sectoral, and temporal aspects of metropolitan change; emphasizes the impacts of long-term change on the metropolitan space economy; examines the implications of his six-phase stages model on the global city-size distribution; describes demographic changes and the role of the manufacturing sector in this process; and correctly argues that urban policy, whether for one city or the entire United States, should be informed by long-term historical trends as well as seen within the global economy context. His approach suggests that not only do economic developers think globally in order to act effectively locally, they also think historically in order to anticipate the future.

5

Regional Theories of Concentration and Diffusion

Regional theories of concentration and diffusion examine economic growth as a process that involves changing industrial structure and spatial structure, both between and within labor market areas. These theories portray the economic growth process as more complex and dynamic than the theories discussed in chapters 3 and 4. They provide a conceptual bridge from early regional theories developed before 1950 to contemporary regional theories developed since the mid-1970s. François Perroux's (1950a) theory of growth poles, Hirschman's (1958) notion of unbalanced growth, Myrdal's (1957) theory of cumulative causation, and Friedmann's (1972) core-periphery model are the best-known regional theories. They are closely related even though they were developed in different contexts or were intended to address different development problems. The common focus is on the issue of whether or not regional disparities in the level of growth and development are likely to remain persistent or even worsen in the absence of public intervention. A major application of the theories—growth center policy—suggests that industrial growth can be diffused to backward regions by concentrating infrastructure investments and direct business investments at selected locations that possess growth potential. Investments in these designated growth centers are used to create spatial industrial complexes from clusters of economically linked industries.[1] Unfortunately, despite the popularity of spatial concentration theories among regional development planners and academics, growth center applications frequently have failed. As we will discuss in greater detail, these failures are due to weak theory and its misapplication.

In general, growth pole, core–periphery, unbalanced growth, and cumulative causation theories move us from strategies focused on a few sectors, factors, or products to strategies that are concerned with detailed economic structure, the interaction between structure and growth, and the spread of growth across space. Although all the theories discussed here emphasize the concentration of growth in particular regions (the polarization of growth) during at least some stage in a development process, their predictions differ as to how the pattern of economic growth across regions will evolve in the long run. Conceptions of the diffusion of regional growth range from Myrdal's hypothesis of persistent regional disparities to Hirschman's belief in strong trickle-down effects and the gradual elimination of disparities.

I. Spatial Concentration Theories

Growth Poles: Perroux's Pure Theory

The term *growth pole* is most commonly associated with Perroux's (1950a, 1950b) hypothesis that growth impulses emanate from particularly powerful actors such as large firms, which operate in an abstractly conceived economic space. Using concepts derived from physics and mathematics, Perroux argued that economic space consists of three principal characteristics: (1) a set of relations between a firm or industry and its buyers and suppliers; (2) a field of forces in which these relations occur; and (3) a homogeneous environment, or "aggregate," in which the forces interact. The concept of economic space as a field of forces gave him the definition of a growth pole: "centres (or poles or foci) from which centrifugal forces emanate and to which centripetal forces are attracted . . . the firm attracts economic elements, supplies and demands, into [its space], or it removes them" (1950a, p. 95). Poles of growth, therefore, represent individual firms, industries, or economic sectors.

Firms, industries, or sectors as growth poles effectively dominate other economic actors through linkages based on commodity and information flows. For example, we can imagine large firms that are able to control markets and dictate terms to dependent suppliers. The "domination effect" exerted by a growth pole may or may not manifest itself in a spatial pattern of polarized growth. Indeed, as Perroux (1950b) argued, domination effects are likely to be nonlocal because the linkages between firms and other economic actors frequently transcend regional boundaries in highly unpredictable ways. A large producer in Atlanta, for example, may be more closely linked to industries in Singapore than to firms in the immediate surrounding regions or states. Perroux actually regarded geographic space as "banal" or unimportant for a true understanding of the economic growth process. He was more concerned that development be studied behaviorally and historically in order to understand the power relations between and among economic actors.

In essence, the theory of growth poles assumes that growth will be an uneven process and addresses how, why, and where growth poles occur. Interested in refuting the tenets of neoclassical equilibrium analysis, Perroux argued that growth proceeds in an unbalanced way, and that changes in the basic structure of the economy as manifested through relations between dominant actors are critical to understanding the development trajectory of particular economies. Using an input-output framework, he sought to explain, via the concept of a "propulsive industry," the structural change that occurs as output increases. Growth poles are propulsive industries that dominate other sectors due to their large size, market power, high growth rate, and high degree of linkage. Subordinate industries, for example, may sell a high proportion of their output to the dominant industry and thus depend on it for competitive success. Growth experienced by the propulsive sector spreads to other industries via input-output multipliers.[2] More importantly, the propulsive sector leads growth by providing a focus for innovation and investment (Thomas 1975).[3] This view is consistent with the ideas of Schumpeter (1934) discussed in chapter 9. According to Perroux, the pole has some edge due to technology, wealth, or political influence, which permits it to increase internal economies of size and scale as well as generate external localization economies for linked industries (Moseley 1974).

Perroux (1988) refined the growth pole concept, seeking to address criticism of his thought, which pointed out the negative consequences of the development of growth poles on the "dominated" units. He argues that growth pole development involves two distinct phases: an initial clustering or attraction phase, in which the dominated units lose resources to the pole, followed by an expansion phase in which goods, investments, and information flow from the pole to the dominated units in order to sustain growth.

It is not difficult to see how these ideas might be applied to geographic space. We can imagine, for example, that a dominant industry's primary linkages could be to firms in the surrounding region, or at least to firms located within the same nation. In either case, the external economies initiated by the propulsive industry affect both spatial relations and economic exchange. Such geographically focused growth might be influenced by regional policy. This logic led many reading Perroux to develop the concept of growth centers, which essentially are propulsive regions rather than propulsive industries.

The fact that so many economic development planners have tried to implement growth pole theory in a geographical context led Perroux to make his own rare foray into spatial analysis. In particular, he articulates a strategy of creating development axes through the establishment of strong transportation links between growth centers. These axes would generate "spinoff development," which would benefit peripheral areas as well. Perroux makes a new distinction between *growth* poles, which conform to his prior definition, and *development* poles, which engender reciprocal economic and spatial relations that can increase overall complexity, "territorialized or non-territorialized." Here, he appears to be arguing that we should first continue to utilize the growth pole theory in terms of economic space and then, by extension and as an outcome of the natural effects of polarization in economic space, apply the theory naturally to territorial space as well. In effect, Perroux is saying that clusters of industries—those linked in economic space by technology and trade—can become industrial complexes through natural co-location. The tendency of firms to be agglomeration-seeking explains this spatial manifestation of economic relations. According to this reasoning, large, dynamic metropolitan areas will dominate smaller cities and their own peripheral areas as they accrue agglomeration economies.[4]

Unbalanced Growth Theory

Although he did not claim to have borrowed Perroux's concept of growth poles, Hirschman was one of the first theorists to describe a development strategy based on the concept of geographical growth centers. In the 1950s, he argued that, in order for an economy to increase income, it must first develop within itself one or several "regional centers of economic strength," termed "growth points" or "growth poles" (Friedmann and Alonso 1964, p. 623). In Hirschman's view, some degree of interregional and international inequality of growth is inevitable—indeed, beneficial. Hirschman terms the spread of growth from the center to peripheral areas—through increases in purchases and investments in underdeveloped areas due to activity in the growth center—the "trickling down of progress." There are also "polarization" effects, such as white-collar migration out of less-developed areas and the dominance of growth center industry. In Hirschman's poor region, the trickling down of progress will outweigh polarization effects; however, "if the market forces that express themselves through the trickling-down and polarization effects result in a temporary victory of the latter, deliberate economic policy will come into play to correct the situation" (Friedmann and Alonso 1964, p. 630).

Hirschman argues that public investment is the policy tool of choice for directing growth. His strategy calls for directing investment toward industries with extensive backward and forward linkages with other industries, which would enable the creation of the most advantageous external economies (economies external to a given firm in a particular industry as well as those between different industries). Because privately financed growth in a given region increases demand for public services such as electricity and water, it is possible for government to induce private investment to a chosen area by installing necessary services beforehand. Dispersal of funds over a large number of projects and among several regions might address concerns for equity and national cohesion but would have little chance of propelling an economy out of stagnation. Public investment would first be used to establish growth centers, then seek to counteract polarization effects generated by the market. Hirschman's ideas were consistent with the views of economic planners in the industrialized and developing world who, in the 1960s, were becoming increasingly concerned that the neoclassical macroeconomic growth models (discussed in chapter 6) were ignoring disparities in regional growth and welfare. Perroux's ideas were becoming more widely disseminated and modified to address regional concerns in the 1960s as well; thus, the growth center concept became extremely influential as a solution to urban–rural disparities, peripheral stagnation, and the growth of mega-cities.

Higgins (1983) argues that growth poles and growth centers became popular because they represented a middle course of action: decentralization with concentration, attention to peripheral areas (equity) with efficient means, and spatial inequality emphasized more than income or wealth inequality. In contrast to economic base theory, which remained popular due to its simplicity, growth pole/center theory attracted a large following because it was highly abstract and vague.

Hirschman's ideas became part of a lively debate in the 1950s regarding the kind of investment pattern likely to be most successful in advancing underdeveloped areas. Like Hirschman, many theorists contended that generating development in stagnant, underdeveloped economies requires focusing private and public investment in just a few key sectors and places to take advantage of economies of scale as well as to maximize the use of scarce financial, physical, and human resources. Other theorists emphasized the lack of effective demand in less-developed regions and countries, a problem essentially ignored by the supply-oriented, unbalanced-growth advocates.[5] These theorists argued for a "big push" of simultaneous investment across a range of complementary industries, to exploit backward and forward linkages and generate external economies. A strategy of simultaneous investment in mutually supporting industries would create and build markets for the economy's production, thus reducing risks and therefore costs.[6] Balanced-growth theorists saw this coordinated wave of investments as a way to overcome underdevelopment, with Nurske (1953) arguing that the very fact that the forces are circular suggests such a strategy can turn cumulative decline into cumulative advance. In this case, the vicious circle becomes beneficent (Rosenstein-Rodan 1943, Streeten 1959, Gianaris 1978).[7]

Proponents of unbalanced growth, on the other hand, saw the larger problem in underdeveloped economies as one of entrepreneurial decision making. From this perspective, development can be spurred by "generating a chain of unbalanced growth sequences in order to induce decision making through tensions and incentives for private entrepreneurs and state planners" (Gianaris 1978, p. 104). Investment in the most strategic sectors of the economy is required to break the vicious circle of poverty. Streeten (1959) argues that bottlenecks created by uneven investment can be a powerful stimulus to the growth of lagging complementary activity. Whereas a balanced investment scheme would lead to the development of plants that would be smaller than those necessary for the optimum use of equipment, initial unbalanced investments in temporarily oversized plants could take advantage of greater cost reductions later (as markets develop). Although overestimates of future demand may lead to losses in the short run, the expansion of capacity is more important than potential periods of slackness.[8]

Cumulative Causation Theories

Gunnar Myrdal (1957) describes a vicious cycle of development in the context of his principle of circular and cumulative causation: economic changes cause supporting social changes in a process that continues in one direction.[9] Because of this circular causation and positive feedback, the dynamic behind poverty in underdeveloped areas becomes cumulative and often gathers speed at an accelerating rate. Myrdal argues, however, that any change induced by organized actors can shift the cumulative process in either a favorable or adverse direction. Moreover, "there is no tendency toward automatic self-stabilization in the social system" (1957, p. 13). Although social forces may array themselves in such a way as to bring a social process to rest, this position is inherently precarious, and any exogenous change may start the process of change in a new direction. Myrdal tries to provide theoretical justification for development process intervention; without intervention, he says, backward areas can be relegated to perpetual underdevelopment, whereas a scheme of balanced locational investment may provide the needed push toward cumulatively positive growth.

Myrdal's analysis of the diffusion of growth is more pessimistic than Hirschman's. The play of market forces can increase inequalities between regions, leaving underdeveloped areas in the "backwater" of growing ones. Less-fortunate regions receive the "backwash" effects of proximity to growth centers, such as outmigration, capital flight, and unfavorable terms of trade since "the movements of labor, capital, goods and services do not by themselves counteract the natural tendency to regional inequality" (1957, p. 26–27). Migration, capital movements, and trade are "the media through which the cumulative process evolves—upward in the lucky regions and downward in the unlucky ones" (1957, pp. 26–27). Essentially, Myrdal is taking issue with neoclassical interregional trade theory (see chapter 7), which stresses how the price mechanism brings the economy into an equilibrium such that disparities between regions are gradually eliminated. Yet, at the same time, he notes that underdeveloped areas may enjoy beneficial "spread effects" from the developed areas with which they are linked, including growing markets for primary goods produced in the poorer area, increasing demand for raw materials (thereby increasing employment and subsequently benefiting consumer goods industries), and the absorption of the poorer region's excess unemployed into the growing one.[10] On the other hand, weak spread effects lead to inequality and ineffective democracy, thus reinforcing the need for activist state intervention.[11]

Myrdal's and Hirschman's counterposed outcomes depend on whether positive feedback (deviation amplifying) or negative feedback (equilibrium tendency) mechanisms are more powerful in the regions under study. Hirschman, like Perroux, recognizes that growth is usually unbalanced initially, occurring in places where innovation and investment are supported; yet he argues that trickle-down effects will ultimately prove stronger than polarization effects. Myrdal sees negative feedback as less powerful than positive feedback; the spread effects promoting development in poor regions will be weaker than the backwash effects that drain resources from these areas.

Friedmann's Core–Periphery Model

Friedmann (1966, 1972) poses a stages-type theory of development based on the notion of cumulative causation. According to Friedmann, the process of economic development involves a critical transition from a preindustrial phase (where agricultural activities are dominant and industry is a relatively small share of economic activity) to a fully industrialized economy. This transition strongly affects the spatial structure of settlement, which, in turn, affects future economic performance. In the preindustrial phase, the economy is dominated by relatively autonomous cities and regions. As industrialization begins, however, and as the economy moves into its transitional phase, investments tend to be concentrated in particular locations, establishing an unequal relationship between these "cores" and peripheral regional areas. The process is actually one of establishing a functioning system of cities.

Friedmann argues that without government intervention to ensure that other communities develop along with core areas, the cores will come to dominate the spatial economy and retard subsequent growth. Writing from the context of underdeveloped countries, Friedmann also is concerned that the poor economic prospects of outlying areas, seen in contrast to the prosperity of core regions, will lead to political instability. Underlying the model, however, is a normative assertion (not necessarily justified with substantive empirical work): a spatial economy focused on a few large urban centers, if allowed to establish itself, will ensure the continued impoverishment of its peripheral areas.

Friedmann's model represents a fusion of regional development ideas with location theory. Indeed, theories of concentration and diffusion are closely related to central place and urban hierarchies theories (see chapter 4), which, essentially, describe the geography of linkages and accommodate the assumed core–periphery or top-down flow of growth. Friedmann (1986) has revised this model in the context of the global economy and the international spatial division of labor.[12]

II. Applications

Growth center policies essentially represent attempts to affect the urban hierarchy as a way to overcome disparities among regions within one nation. One argument is that intermediate-size cities should be favored for development as growth centers because they are large enough to attract capital, yet are sufficiently dispersed spatially to be accessible to commuters from more remote areas. They also offer better quality of life and lower social costs than do large cities. Intermediate cities would offer more opportunities than smaller areas via external economies while permitting fewer diseconomies than larger metropolitan areas.[13] In other words, such intermediate-size centers provide a necessary level of infrastructure and services to serve as central places but have not yet reached the size and scale at which service provision and infrastructure maintenance become problematic, and land prices, other expenses, congestion, and environmental deterioration overtake the benefits to be found there.[14] The Achilles heel of this logical approach pertains to the internalization of externalities. Society absorbs most of the diseconomies of agglomeration. Conversely, private actors enjoy most benefits of agglomeration economies. Although rational government intervention could favor decentralization and growth centers, private investors will still rationally prefer the larger cities.

Growth center strategies were supposed to follow a preexisting pattern of spatial development, to improve on the pattern rather than create a new structure. The ideal spatial pattern would be a dispersed hierarchy of cities, remote enough from each other to be considered "decentralized" and allowing for accessibility to peripheral populations. Although there has never been consensus on the optimal size of a growth center, Hansen (1967) designated a "growth spurt threshold" of between 150,000 and 200,000 people, which represents the ideal range for taking advantage of agglomeration, without the drawbacks of large-city diseconomies.[15]

Although these plausible hypotheses became conventional wisdom in the 1960s and early 1970s, they essentially proved difficult to apply usefully in practice. On the one hand, many policy initiatives proceeded without regard for the types of places most likely to serve as effective growth centers; the temptation to name any needy area a "growth center" proved too great (Higgins 1983). On the other hand, research attempting to establish a relationship between level of economic development and spatial structure could find none. Allan Pred (1976) found that intraorganizational linkages are often more important than spatial proximity (spread from core to hinterland) or urban hierarchy (diffusion from largest centers to smaller ones). Researching growth transmission processes in North America, he asserts that the assumed top-down channels of growth are wrong and that growth transmission occurs primarily between urban centers rather than between each center and its periphery. After citing the role of multilocational firms, which generate significant "non-local multiplier leakages" between urban areas, Pred demonstrates how many urban areas in the United States and Canada lack linkages to smaller urban places but have stronger linkages with New York, San Francisco, Los Angeles, and Chicago than to their own surrounding areas. He then urges planners to focus more on the spatial structure of organizations and on growth transmission studies before adopting intuitive approaches to regional economic development such as growth center strategies. His argument would appear to have greater force in the current era of telecommunications than in the 1970s, when he conducted this research. Returning to Perroux's emphasis on economic over geographic space, he sees corporate organizational structure as a factor too important to ignore in spatial models of innovation and growth.

Although growth center strategies as the primary interregional application of development theories of concentration and diffusion proved flawed, more useful applications of these ideas are available to the economic developer focused on one area. Most metropolitan areas in the United States have experienced a significant growth spurt sometime in their economic history. Often, one dominant industry—one firm, one industrial sector, or one industrial complex—can be identified as the "growth pole" that was largely responsible for the area's growth during that time. Although staple theory offers more useful insights about long-term growth and change, concentration and diffusion theories provide interesting ideas about changes in industrial structure and the spatial pattern of growth.

Another useful idea is the strategy of infrastructure concentration. Whether the economic developer is working at the state or local level with one type of infrastructure (for example, transportation facilities) or with entire planned-unit developments, infrastructure concentration usually represents a more rational allocation of public investment than dispersion. Economies of scale in the provision of public facilities become more feasible. These facilities can also help generate external economies to local firms. Because "pork barrel" politics encourages dispersion, it is helpful to have good theoretical support for concentration. With the growing concern for sustainable development, infrastructure concentration becomes a strategy for preserving or conserving valuable land and natural resources and for utilizing existing capacity, as well as a way to achieve the thresholds required to accelerate economic growth. Economic developers can also use infrastructure concentration, especially in rural areas, as one reason to promote regional cooperation, such as tax base sharing, which is needed yet difficult to achieve.

Finally, the core–periphery model directs attention to interaction within the labor market area between core metropolitan counties and nonmetropolitan hinterland counties. Primary and resource-oriented products in the hinterland may be sold in the core.[16] Commutation flows primarily from hinterland to core. In other words, most nonmetropolitan areas in the United States depend on the adjacent metropolitan center; many residents of the hinterland work in the core area. Their wages and salaries are an important source of purchasing power in the periphery. In rural "bedroom communities," out-commuting is essentially the economic base. Thus, the issue of rural development may be effectively addressed in many such areas through core-hinterland linkages. The economic developer in nonmetropolitan areas dependent on nearby metropolitan areas might use balance of payments accounts, described in chapter 3, to track the interactions. Developers should be able to estimate, at least roughly, the magnitude of inflows and outflows. An assessment based on these flows will suggest whether the interactions are largely beneficial. In other words, the assessment determines whether spread effects or backwash effects are dominant. In either case, developers in these areas need to pay close attention to the plight of the metropolitan center, since the level of local employment responds to jobs provided there.

Growth Poles Application

Growth pole theory reinforces well-known generalities about the economic history of Columbus, Ohio. The original growth pole of Columbus during the nineteenth century was state government. This sector drew labor from the Columbus hinterland and provided services to the entire state of Ohio. Growth stimuli, transmitted through interindustry multipliers, built the nonbasic sector. Ohio State University created additional growth stimuli in the late nineteenth and early twentieth centuries. The Honda plant in Marysville (about a forty-minute drive from Columbus) offers a more recent example of a growth pole. Parts suppliers and service firms have been attracted to the Marysville area to serve the manufacturing facility. This manufacturing growth pole offers a contrast to Columbus's earlier service-sector growth poles. Two other manufacturing sectors based in Columbus—condensed evaporated milk (SIC 2023) and steel foundries (SIC 3324)—are prominent industries nationally and deserve analysis as potential growth poles.

The most straightforward way to apply growth pole theory is to analyze the structure of the Columbus economy in order to identify the dominant industries. Input–output models can be used for this purpose, although the industrial detail must be much greater than the Columbus model presented in chapter 3. Industries with the strongest backward and forward linkages may be dominant industries because they generate the largest local multiplier effects.

Another line of analysis focuses on innovation. It is important to know which firms and sectors are heavily engaged in research and development and in the commercialization of new ideas and technologies. These firms, though not large at present, may have significant growth potential that the economic developer can facilitate. Columbus is rich with university-based and think-tank talent that could become more involved with local, innovative firms. As local linkages and alliances are formed, these firms could well serve as growth poles.

The theories of growth poles and unbalanced growth place great emphasis on economic power. Such power can be manifested in different ways. Local firms may have considerable market power as members of oligopoly industries. Local business leaders may be prominent in national trade associations. Local politics, entertainment, and media may also be strongly influenced by families who own or manage major firms. Given the local power relationships, the economic developer who studies these connections is more likely to understand how to be effective.

Today, Columbus's export sector is diverse, consisting of enterprises and establishments in manufacturing, finance, insurance, retail trade, and distribution, among others. Some companies are headquartered in Columbus, while others are branch facilities. Economic developers would be wise to follow Pred's lead and identify and map the existing pattern of linkages, at least for the most important exporting companies, to determine which are intraorganizational and which are based on spatial proximity. Developers should track the decisions of companies with major branch facilities in Columbus, as well as be aware of important local suppliers and customers of companies headquartered in Columbus. Although it may not necessarily lead to better development strategies, such analysis would elevate the level of understanding of the Columbus economy.

III. Elaboration and Criticisms

The proliferation of regional strategies designed to develop growth centers provides a clear track record with which to evaluate how well growth pole theory and other models of concentration and diffusion have been put into practice. Overall, failures far outweigh successes. Is this the result of misapplication of theory, or are the theories either fundamentally flawed or not well specified? Certainly the vagueness of Perroux's work has caused problems when extending growth pole theory to practice. His noble goal of creating a general theory of economic actors involved in power relations counterposed to neoclassical economics led him to keep the formulation of his thinking at a general level. For example, growth pole refers vaguely to entrepreneurs, firms, and industries, a generality that has invited applications of the concept to widely varying circumstances, often inappropriately. And, despite Perroux's initial warnings not to apply his concepts geographically, they have been so applied many times. Unbalanced growth, cumulative causation, and core–periphery theories are no less general in their specification than growth pole theory.

One of the most basic assumptions of growth center applications has been that the effects of growth will spread or diffuse to the periphery. It is assumed that if the necessary linkages exist, the center's interactions with its periphery will be strong; yet nowhere is it stated how this phenomenon will occur if the linkages to rural areas or slow-growth firms do not exist (Gore 1984). Perroux (1950b) speaks of the domination effect bringing along lagging industries (later applied to lagging peripheries). Yet he does not seem to question that this outcome might not occur in all circumstances. It is useful to recall the context in which Perroux first formulated his growth pole theory. He was writing in Europe during the early 1950s, a time of relatively primitive communication networks between regions and nations when peripheral regions, abundant in vital natural resources, could directly experience spread effects from nearby centers. For example, areas of iron, coal, or forest reserves were directly affected by the growth of nearby mill towns. Despite his general avoidance of geographic applications of his theory, Perroux was undoubtedly influenced by this European context.

It appears that as long as there are strong backward linkages between an industry in a growth center and its natural resources at the periphery, spread effects can be felt in that periphery. Once industries become human resource-based, however, these spatial links are broken. With this insight, Higgins (1983) proposes a stages theory of growth poles: a process of change from a propulsive region of primary (natural resource) production, which causes growth to occur at the center; to a propulsive center based on primary product processing, which causes growth at the periphery; to a final stage of industry at the center with more connections to other centers than to its own periphery. Thus, growth poles as a concept may be relevant only as a spread mechanism in the second stage. It is worthwhile to look at growth poles in this temporal context, especially when one remembers that Perroux himself urged taking a dynamic approach to growth and development.

An outcome of the vagueness of concentration and diffusion theories has been their almost indiscriminate application to regions of various sizes. In the United States, the Economic Development Administration (EDA) designated small, stagnating rural towns of 500 or so people as official "growth centers," to be assisted by funding and investments in propulsive industries. This approach appears to have failed miserably. It seems that, in such cases, policymakers mistook the *need* for growth and development in a region for the *potential* for such growth and development. In trying to meet both efficiency and equity goals, they sacrificed the former to the latter. As a result, large amounts of funding flowed to these depressed areas, but little economic growth resulted. Although the EDA has long abandoned this program, states and local jurisdictions continue to misapply growth-pole logic. They provide incentives or focus infrastructure development in peripheral rural areas in the hope of attracting private investment. Yet this supply-side approach has rarely reversed the fortunes of nonmetropolitan areas in need.

Growth center strategies also tended to emphasize "once-over actions," or an initial outlay of capital, to stimulate the growth that becomes self-sustaining through the domination effect. Perhaps too much emphasis has been placed on Hirschman's unbalanced growth thesis, although Hirschman himself would have stressed a sustained investment effort. This "one-shot" emphasis led, in part, to an inappropriate bias toward investment in industrial infrastructure—for example, industrial parks and "spec" buildings, at the expense of continual investment in educational and social services. A more self-sustaining approach to growth center development would be to focus on the ongoing development of human resources and local administrative capacity, and on the decentralization of political power (Moseley 1974).

For the small growth center, a heavy investment in one type of industrial infrastructure may prevent adaptation to changes in macroeconomic climate and product cycles without further subsidy.[17] In addition, investments in heavy industry, to generate "propulsive effects" on a developing economy with a large rural sector, often miss the mark. The existing economic activity in many medium-size centers in developing countries focuses on the processing and marketing of agricultural products. Much of the marketing activity is small-scale and is considered to be part of the "informal" economic sector—that is, it is nonregulated and nonstationary and requires little capital investment and overhead costs. Such informal marketing activities often generate relatively large amounts of income and employment. It follows that, in such a context, a growth pole strategy would be best focused on harnessing the entrepreneurial drive that undergirds the informal sector. Investments in agro-processing, basic marketplace infrastructure, transportation links, and credit extension would thus be appropriate. In the early 1980s, secondary city development was advocated as a way to concentrate investments in these areas to spur local economic development appropriate to the local economic base and thus diffuse benefits to the periphery. Consensus on this approach never existed, however, and by the 1990s lending institutions had largely abandoned such efforts in favor of macroeconomic approaches, such as structural adjustment.

Another criticism of growth pole theory is that in today's world, it is focused at the wrong level—the national level. Transportation and communications linkages are transforming the structure of economies around the world. If Pred's work were extended internationally today, many primate cities around the world would have stronger economic linkages to other large cities elsewhere in the world than they would with cities in their own country. Especially true for those "poles" specialized in the services sector, this also is true for manufacturing complexes. Spread effects to a center's periphery and to smaller centers may be weaker now than ever before.

At the same time, the local level has become more important; local labor market areas are now the functioning economic unit in the global economy. Local economic developers have become more relevant actors as the overall influence of macroeconomic policy has waned.[18] At the very least, these developments mean that a close examination and adjustment of theoretical assumptions must occur before any spatial economic development strategy, such as growth centers, is embarked upon.

Economic developers have been preoccupied with seeing growth pole theory as a basis for growth center strategy, rather than as a way to understand economic development. Thus we have seen growth center designation in clearly stagnating areas, large-scale industrial projects in isolated rural areas, and other misallocations of investment capital. We have also seen blanket applications of the growth center strategy to widely varying contexts without consideration of the "stages" of growth pole development that make certain contexts clearly inappropriate for such a strategy. It is safe to say that growth centers are no longer useful in the context of the developed, postindustrial economies found in North America, where linkages are often human resource-based and different developmental needs arise. Multilocational firms play an important role in strengthening the linkages between headquarters and branch locations at the expense of linkages between the firm and its geographic periphery. These firms can have detrimental effects on developing regions when they fail to generate significant spread effects to peripheral areas, and they can actually aggravate rural-urban disparities.

In conclusion, theories of concentration and diffusion are not self-contained sets of rules, principles, or statements designed to explain a phenomenon. Many of these ideas are essentially descriptive; to gather explanatory power, they must be elaborated and carefully specified. Perroux's original work often seems as though it was co-written by an economist and an amateur physicist struggling over some grand professional compromise, whereas the refiners of Perroux subsequently accumulated baggage from an assortment of economic and spatial theories such as location theory, central place theory, and agglomeration economies. All this was done with the intention of both clarifying Perroux's ideas about economic development and attempting to put them into practice as a normative development strategy. The result of implementing a concept without its own unified theoretical base has been the misapplication of Perroux's ideas, depending on the "spin" of the practitioner. An updated and supportive treatment of Perroux's theory is found in Higgins and Savoie (1995).

Higgins states that we need more knowledge before we can apply growth pole theory effectively. He emphasizes the need for better understanding of growth and change in different settings, and of the role of market forces in creating and alleviating regional and income disparities. He also calls for examining the potential of different types of government intervention in guiding such market forces. Higgins advocates research into the role distance plays today in all these processes. Given the ongoing revolution in telecommunications, this aspect takes on even greater importance. Finally, he recommends research into city hierarchies and sizes, and the role they play in development.

Higgins's admonitions would also apply to the other related models—unbalanced growth, cumulative causation, and core-periphery. Unbalanced growth has been both updated and subsumed in neoclassical growth theory, especially recent specifications that treat spatial factors more explicitly (see chapter 6). Cumulative causation is addressed more fully from the perspectives of growth theory and trade theory (see chapters 6 and 7). The revised and updated core-periphery model is now specified in the context of world-city research (Friedmann 1995).

Discussion Questions

1. Which ideas from growth pole theory appear to be most applicable currently to metropolitan economies in the United States?
2. Why did growth center strategy become so politically popular? What can we learn from the failures of this strategy?
3. Can you construct an argument to defend the concentrated provision of public infrastructure as a means of making development more sustainable?
4. Are the linkages of major industries in your area to other industries within the region, or are they connected to industries located elsewhere?
5. Why would the unbalanced growth approach stimulate more entrepreneurship than the balanced growth approach? Which should be more effective in impacting the "vicious circle" of underdevelopment?
6. Under what circumstances should spread effects be more powerful than backwash effects?
7. Are there important differences between the diffusion of growth from more-developed to less-developed countries, compared to the interregional diffusion of growth?
8. How could increasing returns (internal or external) dampen the negative backwash effects?
9. Is the application of cumulative causation to racial relations and to regional development a strength or a weakness?
10. Are agglomeration economies important to all local firms? Are there major local companies that rely more heavily on external linkages?
11. Each concentration and diffusion theory appears to have an implicit political theory. What is the role of local government? In general, what is the role of the economic developer?

Notes

1. Attempts to identify or create promising growth centers have led to important contributions to the literature on central place theory and urban hierarchy. See chapter 4.

2. In a major review of the growth pole literature, Darwent (1969) describes "propulsive industries" as having distinct characteristics: high interaction with other firms (that is, linkages), a high degree of dominance, and greater than average size (to generate economies of scale and economies internal to the industry). The linkages can be either *backward*, with other firms involved in supplying factors of production or intermediate inputs, or *forward*, relating to all sectors other than final

demand. The strength and importance of these linkages determines a propulsive industry's role as a growth pole. Interindustry flows can be used to assess the relative importance of inputs and outputs and, therefore, the potential dominance of one industry over another. See the discussion of input-output analysis and linkages in chapters 3 and 4.

3. Perroux's ideas regarding the role of the innovator or propulsive industry can be seen as a mediated approach to economic theory. Like Schumpeter, he explicitly elevates the importance of the role of individual innovators and their decision making in the economic process. Thus, his "active units" involve human agency in creating an engine of economic growth.

4. Amos (1990) views development as a global process governed by long cycles of about a hundred years. Spatially, growth poles become dominant and polarization occurs during half the cycle, then these poles diffuse growth and development over the next fifty years. He speculates that, around 1980, the diffusion cycle from U.S.-based growth poles was ending and a new concentration phase was beginning, led by growth poles arising in Japan and in other Pacific Rim countries. Amos's article includes concise summaries of growth pole theory and long-wave theory in presenting a spatial-temporal synthesis similar to that of Suarez-Villa (1988).

5. Underlying balanced growth arguments is the notion that certain socioeconomic forces act to keep less-developed economies or regions in a permanent state of underdevelopment. The problem may not only be lack of capital, but the lack of demand for capital as well. On the demand side, it may be that the small size of the market in an underdeveloped economy dampens any inducement to invest. The market size is determined by the buying power of the populace, which is limited by low real incomes, which are in turn a result of low productivity. Productivity remains low due to minimal inducement for investment, thus a "vicious circle." On the supply side, low per capita income fails to satisfy the needs of consumption and so effectively limits the capacity to save. But low incomes are a function of low productivity, and productivity is dependent on the capital investment that is itself limited by low savings.

6. A scheme of planned industrialization, as envisioned by balanced growth theorists, necessarily requires substantial state intervention in the workings of the market economy. Firms are concerned with what is profitable in terms of private net marginal product and not with social net marginal product. The latter, which would include the benefits derived from pecuniary external economies, should ideally be considered in any calculus of profitability. If investment is based solely on potential profits for private firms, inducement to invest in any given project will be inadequate when rational investors, acting under profit motives and in isolation, perceive market deficiencies. Although the marginal productivity of capital over a range of complementary industries (encompassing many individual projects that such investors would avoid) is often considerable, rational individuals would fail to recognize this potential. One role for the state, then, is to ensure that producers do not subjectively overestimate the risks of production and underestimate potential benefits.

7. Though much of the literature regarding the balanced/unbalanced growth argument was written during and after the 1950s, previous authors dealt with similar issues. Friedreich List advocated balanced investment among agriculture, manufac-

turing, and commerce, and argued that it was the duty of the state to bring such a policy into existence (Gianaris 1978). Conversely, Adam Smith and the Physiocrats stressed growth in the agricultural sector as a means of creating surplus food production to feed growing urban populations. Thomas Malthus argued that increasing returns justified concentrating investment on industry.

8. Have countries and regions that experience faster rates of growth developed in a balanced or unbalanced fashion? In a series of tests using different combinations of sectors for a sample of forty developing and developed countries over the period 1938–1960, Swamy (1967) generally found positive and statistically significant correlations between measures of sectoral variability and aggregate economic growth rates. The results proved robust across different time periods and in an analysis of countries at different levels of development. He concluded, somewhat cautiously, that the evidence "points out that sectoral growth rate imbalance need not inhibit the overall growth of the economy" (1967, p. 300). Demery and Demery (1970, 1973) also published evidence of a positive relationship between unbalanced growth and aggregate growth rates. Tests by Youtopolous and Lau (1970), however, indicated that a high level of variation in sectoral growth rates tends to be negatively correlated with overall growth rates. Their analysis suggests that the fastest-growing places grew in a balanced fashion.

9. Myrdal first challenged the notion of the equilibrium view of the social process in his 1944 study, *An American Dilemma, The Negro Problem and Modern Democracy*. He derided the prevailing view that the level of social and economic development of the black population would necessarily be slow and uncertain, that any state intervention would be ineffectual at best and counterproductive at worst. The status of the American black, which seemed to be stagnating at the time Myrdal wrote, suggested the evolvement of some sort of "stable power equilibrium." Myrdal predicted that this equilibrium was only a "temporary interregnum." He was seemingly proven correct by the subsequent upheaval in interracial relations that occurred in the 1960s and 1970s.

10. The concept of spread-backwash has been subjected to a number of tests using several different methodologies, including spatial regression methods (Casetti, King, and Odland 1971; Lewis and Prescott 1972) and trend surface analysis (Robinson and Salih 1971). In his analysis of twenty-one studies of spread-backwash, Gaile (1980) found that although promising methodological work had been done, not only had the empirical findings of the work been inconclusive, but the research in general had neglected a "theoretical focus." Most of it also failed to examine spatial unbalanced/balanced growth in an interregional context, concentrating instead on the intraregional scale of development. This focus is noteworthy, since the original concept, as formulated by Hirschman and Myrdal, centered on interregional development.

11. Unfortunately, because spread effects in less-developed countries are characteristically weak, the low level of development, coupled with regional inequalities, combine in the cumulative process such that "poverty becomes its own cause" (Myrdal 1957, p. 34). This process suggests that the state must promote policies favoring underdeveloped regions. Myrdal notes that advanced European countries have approached the status of welfare states precisely because stronger spread effects caused less inequality, a social luxury that provided for the growth of effective de-

mocracy. These states are characterized by a "quiet contentedness" that is the result of national integration achieved through government regulation of market forces.

12. Given his understanding of the spatial division of labor in the global economy, Friedmann (1986) has revised and updated this model. He argues that world cities now constitute the core areas of the economy, whereas underdeveloped regions comprise the periphery. The core areas fulfill command and control functions and are locations of capital accumulation. They also serve as places where migrants are destined, which generates high social costs. The revised model, grounded in neo-Marxist theory, emphasizes capitalist contradictions and the resulting polarization between core and peripheral countries, between regions within countries, and within metropolitan areas. Friedmann synthesizes world-city research as pertaining to (1) large nodal regions (urban core areas) that embody intensive interaction, (2) a changing global urban hierarchy based on command and control, (3) capital accumulation predominantly in core areas, and (4) a powerful "transnational capitalist class" that is engaged in conflict with more localized classes (1995, p. 26). Appendix 9.1 provides an overview of Marxist theory.

13. Agglomeration economies figure prominently in the reasoning of growth center advocates. As in growth pole theory, agglomeration involves maximizing interindustry linkages and multiplier effects. However, unlike growth poles, growth centers focus on encouraging the organization of these linkages in geographic proximity. Clustered public investment in services and infrastructure encourages agglomeration. The agglomeration of industrial activity results in cost savings as firms share the benefits of infrastructure, amenities, and the exchange of information that result from close proximity.

14. In the context of developing countries, Rondinelli (1983, p. 14) also called for such a targeting of intermediate-size cities, arguing:

> they offer economies of scale for a wide variety of basic social and economic activities, organize the economies of their hinterlands, provide access to transportation and communications networks, offer off-farm employment in tertiary or secondary sectors, and provide access to markets, services and facilities in larger towns and cities.

An integrated system of such cities would spread the benefits of growth and innovation from the primate city to other areas. He cited the lack of such a hierarchy or system of central places as a reason for the failure of development to diffuse to nonprimate areas.

15. The issue of optimum city size has a long history in urban planning and regional policy. Planners focused on minimizing the average cost of municipal services argue for cities in the 200,000 to 500,000 size range and against large cities. Alonso pointed out that marginal revenue and marginal cost may be more pertinent than focusing solely on average cost. These marginal rates may be equal at city sizes well above 500,000 or 1,000,000 if average and marginal revenue increase with city size. In advocating intermediate-size cities around 200,000 as growth centers, Hansen (1967) implies that average revenue does not increase monotonically with size.

16. Most U.S. territory is in rural or nonmetropolitan areas. The economic base of some of these areas is in primary products where agriculture, forestry, fishing, mining, and oil and gas extraction dominate the local economy. To these places one can

add places dominated by outdoor recreation and related tourism services. Parts of the western United States, for example, have economies that are driven by the national parks in the region. Other relatively isolated rural communities have grown as retirement communities. As the "baby boomers" enter their retirement years, these communities will become more important. Such areas rise and fall with the economic base/export staple in ways described by these theories in chapters 3 and 4.

17. In other words, overemphasizing efficient production can prevent the development of a region's ability to produce flexibly or to innovate adaptively in response to changing markets and "niche-filling" opportunities. See chapter 11 for further discussion.

18. In the 1990s, the economic power of regions has increased, as has the political power of local governments. For further discussion, see Malecki (1997) and Ohmae (1995).

6

Regional Growth Theory

All the theories discussed in this book have to do with economic growth in some form. Two characteristics distinguish the theories outlined in this chapter—and their classification under the heading "regional growth theory"—from others treated elsewhere. First, these models are concerned primarily with explaining movements in a fairly narrow set of key macroeconomic indicators, such as output, employment, income, investment, savings, wages, and interest rates. The other regional development theories, with the exception of economic base, to the degree that they address the issue of growth, generally explore a wider range of economic and social issues both at the macro and micro economic levels (for example, growth poles with their emphasis on interindustry linkages and entrepreneurship with its focus on the behavior of key economic agents). Economic base theory is, in fact, a Keynesian model of income determination—definitely within the purview of models described here—although it is usually considered separately (as in chapter 3).

The second distinguishing characteristic of the theories presented in this chapter is that they are most relevant to understanding *interregional* growth and development trends. Here, we abandon the implicit one-region/rest-of-world perspective of chapters 3-5. Theories of regional growth address this question: What determines observed disparities in regional income and income growth? An important related question concerns whether disparities are likely to persist or whether they will disappear through the natural play of market forces. Several models suggest that the former is true, and that there is a potential role for government intervention in alleviating disparities. Other theories predict that eventually incomes across regions will converge, and therefore that regional development policies are unnecessary, even wasteful. Because existing empirical work is only a partial guide, it varies in its conclusions (Chisholm 1990). Understanding regional growth theory is important for local economic developers, if only so they can comprehend why some citizens, businesspeople, and elected officials remain skeptical about the need for publicly funded development initiatives.

This chapter outlines two bodies of regional growth theory: (1) neoclassical models, including recent revisions described as endogenous or new growth theory; and (2) what we will term, following McCombie's (1988a, 1988b) example, "post-Keynesian" models of regional growth. At the risk of oversimplifying, we can say that both traditional and new neoclassical growth theories focus on the problem of supply. That is, assuming sufficient demand, what prevents the economy from achieving its maximum rate of growth? The issue, from the neoclassicist's point of view, is a matter of determining the factors that prevent firms from producing at full capacity, such that economy-wide resources are allocated most efficiently. Post-Keynesian growth theories generally focus on deficiencies in demand. Specifically, for what reasons is demand insufficient to fuel additional production and growth in some regions? Local economic developers encounter these supply-side versus demand-side perspectives more frequently than they probably realize.

Consider the case of the potential impact of unionization on the local economy. On the one hand, any wage increases due to unionization may have a dampening effect on local industries by directly increasing costs relative to those at other (nonunionized) locations. On the other hand, workers earning a pay increase have more money to spend. Consequently, they demand more goods, potentially fueling additional regional economic growth. Endogenous growth theory, also referred to as new growth theory, represents an extension of the neoclassical model to account for increasing returns to scale and externalities. Interestingly, endogenous growth theory is capable of generating conclusions that closely mirror some of those of the post-Keynesian approach, particularly cumulative causation theories. Thus, the two perspectives on regional growth have, to some degree, moved closer together at least in terms of some of their conclusions.[1]

I. Overview

Like many models of regional development that have gained influence in the United States, regional growth theory is dominated by the neoclassical perspective (see Appendix 6.1 for a brief discussion of the neoclassical paradigm). Neoclassical growth theory focuses principally on supply-side issues. The nature and magnitude of demand for goods and services typically are ignored when they are not regarded as irrelevant. Although early neoclassical growth models de-emphasize or discourage public-sector development intervention in the market, the new (neoclassical) growth theory admits a more significant role for economic developers in encouraging growth or reducing regional disparities. Indeed, the supply-side focus of much U.S. economic development activity—infrastructure development, training programs, technical assistance, and science and technology programs—may find some justification from new growth models that explain regional industrial specializations in terms of externalities and agglomeration economies. With their emphasis on insufficient demand for the output of regional industries, post-Keynesian theories have had a more limited, though still important, influence on regional development thinking. Their impact on policymaking has been stronger in Europe than in the United States, partly because they imply a need for a much stronger and explicit regional policy framework than is politically palatable in the United States.

Neoclassical Theories of Regional Growth

Using a concise set of economic maxims as a basis for analysis, the single-sector neoclassical regional growth model postulates that the process of economic growth within a nation leads to reduction, then elimination, of productivity growth and per capita income disparities across regions. Because the system tends to achieve and maintain this equilibrium result on its own, there is no need for development policies to encourage growth in regions that are underdeveloped or lagging. In fact, government intervention is likely to do more harm than good, since the market, through the price mechanism, determines the most efficient allocation of resources. Thus, reduction of allocative inefficiencies in the market (often caused by existing government policies) constitutes the theory's key policy recommendation. More complicated multisector, neoclassical growth models yield less precise conclusions, with some models even predicting that regional disparities may persist. Nevertheless, the potential harm from government action is widely regarded as likely to be greater than any benefit derived from intervening in the workings of the market.

Despite the prominence of both supply and demand in any treatment of elementary economics, the difficulties associated with accounting for both sides of the market as models become more complex mean that many development theories restrict their attention to only one side of the regional economy. According to economic base theory, for example, any change in local output is determined by increases or decreases in the demand for regional exports. There are no supply constraints; presumably, any increases in demand can be met with concomitant increases in production. In contrast, neoclassical regional growth theory focuses exclusively on the role of supply as the fundamental determinant of changes in local labor market or metropolitan area output.[2] Here demand is assumed to be perfectly inelastic[3]; output growth is determined by growth in regional factor supplies and technological change. In fact, the basic neoclassical growth model implicitly invokes Say's Law: supply creates its own demand (McCombie 1988b).

Neoclassical *regional* growth theory may be described as a "regionalization" of models of national economic growth originally developed in the 1940s and 1950s. National-level neoclassical growth models focus on the economy's long-run *potential* growth path. At any given time, the actual level of production in a given economy may deviate from its potential level, where the latter represents a state of full employment. Because a literally zero rate of unemployment is never possible, potential output, defined more precisely, is the level of output that can be achieved at the lowest rate of unemployment consistent with no inflation.[4] Neoclassical growth theorists assume that short-term government stabilization measures (monetary and fiscal policy) maintain the economy at full employment. They then attempt to determine whether the economic system tends to grow at what is referred to as a steady-state, stable rate of growth. The simplest growth model suggests that the rate of long-run, potential growth is determined by technological change and the growth in the size of the labor force through natural population increase.

The simplest neoclassical model of regional growth incorporates the unique features of subnational geographic units. This means that, whereas the model's basic framework and assumptions are identical to the national model, it accounts for the possibility that labor and capital will migrate between regions by assuming that both factors will seek the locations that offer the best returns. Workers will be attracted to high-wage regions, while capital will migrate to regions offering the highest rates of return. The model also assumes that the latest technology is available to all producers, regardless of location. This implies the instantaneous diffusion of productive innovations over geographic space; that is, once an innovation is developed, there are no barriers preventing particular firms (perhaps those located in outlying or rural areas, for example) from taking advantage of the innovation.

From the traditional neoclassical perspective, the process of regional growth may be described as follows (see Appendix 6.2 for a detailed exposition of a simple neoclassical growth model). Assuming a constant returns production function, the level of capital stock, labor, and technology in the first time period determine, through the production function, the level of output in the same period. In turn, output growth is determined by the growth in factor supplies and technology. Because technology is assumed identical in all regions, differential rates of growth across regions in any given period may be due to differences in indigenous labor force growth and factor migration. In the long run, the system tends toward an equilibrium where productivity growth, wages, and the rates of return to capital are equalized across all regions; as in the national case, long-run growth is determined by growth in technology and the natural rate of increase of the labor force.

The findings with respect to regional disparities hinge on the migration and returns to scale assumptions. Faster-growing areas, for example, will not enjoy persistently higher wages than those in slow-growth regions, since their attendant higher rate of labor force growth will dampen wage rates. The same dynamic applies to returns to capital. At the same time, the rate of growth of output across regions may vary, depending on the natural increase of each region's indigenous labor force. Moreover, it is possible that some faster-growing regions may have to import capital from slow-growth regions in order to equalize saving and investment. As a consequence, some regions may import, and others export, capital. Interregional capital flows will be observed even though the rate of return everywhere is equalized (Borts and Stein 1964; McCombie 1988a, 1988b).

Figure 6.1 summarizes the determinants of the simple neoclassical regional growth model.[5] The automatic equilibrating mechanisms of the model imply that there is little need for government intervention to encourage regional growth in depressed places. Over time, regional output growth disparities will narrow and eventually converge; productivity growth, and thus growth in regional wage rates, will be determined by the exogenous rate of technological progress. Any policy attempts to accelerate this process will likely generate inefficiencies. As we have noted, the model offers few policy recommendations for the development practitioner, short of minimizing (indeed, eliminating) government interference in the market.

Thus far, the results are based on an assumption that the local economy consists of a single sector. This particular formulation is often referred to as the "naive" model of neoclassical growth (Richardson 1978). Because neoclassical theorists themselves have extended the analysis to multiple sectors, this chapter's characterization of the one-sector model as the fundamental neoclassical approach may be somewhat unfair.[6] As soon as multiple sectors are introduced, the model admits the possibility that labor and capital flow in the same direction, perhaps leading to persistent disparities in growth rates between regions. This possibility reopens the door for renewed speculation regarding whether certain cities and regions are able to consistently outcompete other areas, not to mention the fact that it yields a much more varied array of implications for economic development practice. Depending on how the multisector model is specified, local practice that encourages

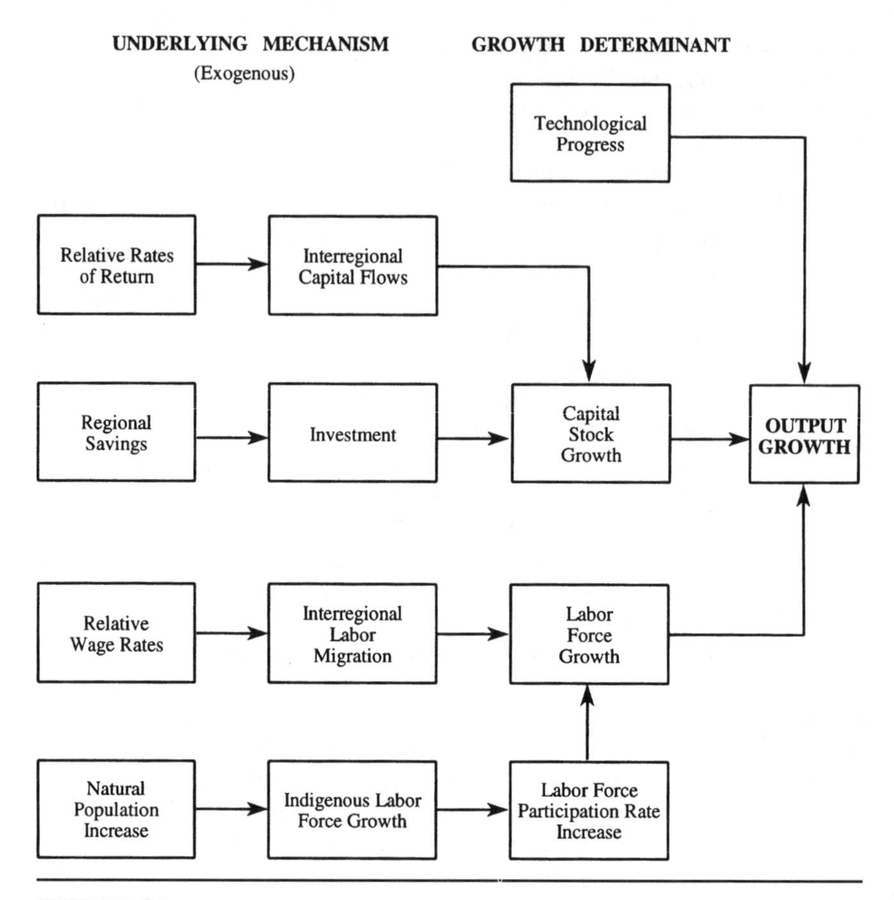

FIGURE 6.1
Determinants of Regional Growth

the growth of the export sector may, indeed, lead to increases in regional growth. Alternatively, if the importance of demand is assumed away, local policies designed to reduce the inefficient allocation of resources between sectors in a given region become important.

The neoclassical growth model suggests that output growth in the near term is determined by growth in the capital-labor ratio. Investments in capital increase productivity; with improved facilities incorporating the latest technology, as well as additional, better machinery and equipment, the same number of workers can produce more and more goods. However, if we assume that the size of the labor force remains steady and that technology is constant, it becomes clear that output growth cannot continue indefinitely. That is because capital, like labor, is subject to diminishing returns. At some point, the additional increment of capital added to an existing workforce can yield no productivity improvement. Just as adding workers to a plant of a given size eventually will lead to overcrowding and a fall in productivity, capital can also overcrowd the production process. At the same time, it is the prospect of achieving returns that induces investment in capital. Therefore, the neoclassical model implies that when capital reaches the point where diminishing returns set in, investment will cease and growth will come to a halt. This is why, in the long run, exogenous technological progress and/or increases in the size of the labor force are needed to sustain growth.

In the traditional neoclassical model, the factors driving long-run growth are therefore exogenous, or outside the model itself. Endogenous growth theory constitutes the latest work in neoclassical growth economics; indeed, it represents a revival of growth economics, an area of research that had essentially stagnated in the 1970s (Grossman 1996). One of the most important contributors is Romer (1986), who initially developed a model demonstrating that long-run growth was possible even with no technological change and with diminishing returns to capital. Romer's innovation was to combine a neoclassical production function subject to constant returns with externalities or social increasing returns. Social increasing returns are essentially cost savings that accrue to firms as the number of businesses or overall size of the industry increases. Romer argued that capital investments generate externalities such that all firms together do not face diminishing returns. Called spillovers, the externalities are related to the growth of knowledge (Romer's model) or to human capital (Lucas 1988).

The importance of endogenous growth theory for regional analysis is that it highlights the important role of social increasing returns and their potential spatial dimension (see, for example, Palivos and Wang 1996, Ioannides 1994). Although most growth economists refer to spillovers as externalities that accompany economy-wide advances in knowledge and skills, there is reason to expect some of these effects to be localized. Indeed, to a large extent, local effects are consistent with the theories of agglomeration economies and cumulative causation, a point noted by some economists, such as Krugman (1991, 1997). Lucas (1988), for example, argues that, to be useful for further theoretical and empirical work, the external effects of human capital or knowledge need to be specified in greater detail. Lucas's external effects of human capital, for example, have to do with influences economic actors have on the productivity of one another. The scope of such effects depends on the "ways various groups of people interact" (1988, p. 37). In particular, these effects could conceivably be regarded as either global in nature or purely localized at the level of family or firm. There is likely some middle (geographic) ground, however, since both individuals and firms typically interact at a larger social scale—that is, the community or neighborhood, city, and industrial district or complex. The cost savings and enhancements in productivity that a firm gains by locating in proximity to other firms in given cities and regions become a source of long-run growth for those places. Endogenous growth theory suggests that if we want to understand why some cities and regions grow, while others stagnate or fall behind, we need to look closely at the nature of these external benefits, the way in which they are encouraged, and how they are inhibited.

Post-Keynesian Regional Growth Theories

In contrast to neoclassical regional growth theory, post-Keynesian models of regional growth emphasize the lack of equilibrium in the growth process, the dependence of local fortunes on the strength of effective demand for regional exports, and the tendency for growth trends to become cumulative (in either a negative or a positive direction). The approach is based heavily on the Keynesian view of the economic system (although Keynes himself wrote little on regional problems). McCombie (1988a, 1988b)—who provides a useful synthesis of the theory, along with a comparison to the neoclassical approach—adopts the term *post-Keynesian* to distinguish this perspective from the "Keynesian-neoclassical synthesis," which, according to Dome, is "the synthesis between Keynes's *General Theory* and neoclassical economics; between macroeconomics and microeconomics; and between a fiscal and monetary policy and laissez faire" (1994, p. 245). An example of the Keynesian-neoclassical synthesis is the (neoclassical) regional growth theory outlined above, with its tepid acceptance of the possible need for demand management strategies in the short run to maintain full employment, but the belief that price adjustments would work in the long run to eliminate regional income and productivity growth differentials. Pinning down the slippery terminology is less critical than understanding the fundamental differences between perspectives. Perhaps most important is that the post-Keynesian approach does not subscribe to the view that markets are generally equilibrating; thus, the theory admits a much broader scope for government intervention.

McCombie argues that the post-Keynesian perspective "provides a unifying theoretical basis for the export-base (or export-led growth) theory and the cumulative causation and polarization theories of Myrdal (1957) and Hirschman (1958)" (1988b, p. 400). This is because the theory asserts the primacy of the export sector as well as the tendency for regional growth in output to lead to increases in productivity through the dynamic of internal and external returns to scale. (Note that traditional neoclassical models as-

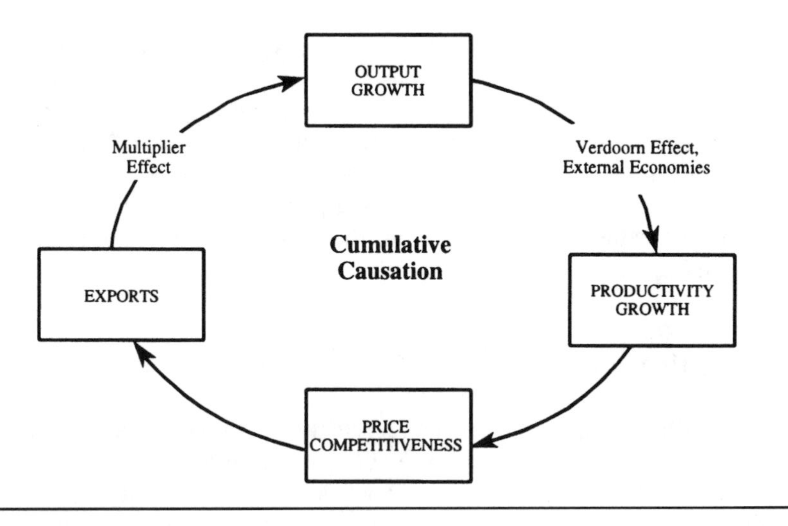

FIGURE 6.2
Simplified Post-Keynesian Regional Growth Model

sume constant returns to scale.) A simplified view of the regional growth process according to the post-Keynesian view is provided in figure 6.2. Output growth by producers in a given location drives increases in productivity through returns to scale; improvements in productivity make the export sector of a region more price competitive vis-à-vis producers in other locations; price competitiveness stimulates growth in exports as consumers elsewhere buy more of the region's goods; purchases of regional exports generate further growth in regional output through a multiplier effect (McCombie 1988b). The linkage between output growth and productivity growth is known as the "Verdoorn effect." Growth of the region's production stimulates an influx of both workers and investment (via migration); the growth of productive factors is not the cause, but the consequence, of output growth. The entire process can be viewed as a formalization of Myrdal's (1957) notion of cumulative causation outlined in chapter 5, and it can characterize a process of cumulative decline as well as cumulative advance.

The question for the regional economic developer is how output growth can be stimulated by increases in exports. That is, how can the performance of the export sector be improved? Several factors are important, including prices, nonprice aspects of competition, income elasticity of demand for the region's exports, and the region's income elasticity of demand for imports (Thirlwall 1979, 1980; McCombie 1988b). To the degree that prices are set at the national level, the region's industrial structure (for example, the concentration of economic activity in industries producing income-elastic goods) and such factors as product quality and service will determine its export prospects. The potential for some regions to gain a sustainable competitive edge in the production of some goods through increasing returns, externalities, and agglomeration economies means that other regions may fall progressively further behind, or at least face below-average rates of growth for sustained periods. In fact, post-Keynesians would argue that, in a world of increasing returns, the tendency of some regions to lag behind others is the rule rather than the exception.

II. Applications

Early neoclassical growth theory suggests that economic developers should do what is necessary to support the efficient allocation of resources and the operation of the price mechanism. For example, they might attempt to reduce aggregate inefficiencies in the economy by helping inefficient companies find more profitable locations or unemployed workers relocate (essentially, both are ways to assist in the economic adjustment process). Such assistance should take the form of providing information to companies and workers, rather than direct financial or in-kind support. Clearly, most economic developers would not be comfortable with such a restrictive role. Modifications and extensions of this simple model, however, open up numerous avenues for intervention—for example, strategies for skills training, industry networks, and expanding existing industries become relevant.

Neoclassical growth theory applications have been much more common in the developing-country context where the monetary economy and market mechanisms are much weaker than in the United States. Many U.S. cities and communities are parts of industrial regions that function as integral elements in an increasingly sophisticated and open global economy. Because these places are integrated into that global system, proper application of neoclassical growth theory suggests that economic developers should rely heavily on market mechanisms and information flows, rather than devise forms of government intervention. Indeed, developers should focus on activities consistent with the functions of traditional local government. Even in the competitive market economy, public goods and services must be collectively provided and financed. The objective is to provide needed public goods and services as efficiently as possible, thereby minimizing the local tax burden on companies and workers. This efficient-government strategy has another benefit, in that it would potentially increase the amount of business and personal savings available locally. More local savings could lead to more local investment as long as competitive returns were available in the region. Although most economic developers are not interested in functioning as city or county managers, they still need to forge good working relationships with these managers because of the value of sound public-sector management.

Another application consistent with neoclassical growth theory involves the analysis of existing forms of government intervention in business activities to see which economic development activities underway actually are needed and which are counterproductive. Which local economic problems could be better solved by the private sector? Are the costs of intervention less than the benefits sought? Though certain to be unpopular with traditional economic developers, the elimination of some common but unnecessary economic development practices may improve the allocation of what are usually scarce local resources.

From the post-Keynesian perspective, a strong case can be made for a regional policy at the national level. Development incentives and infrastructure investments to lagging regions may help reverse a process of cumulative decline. Currently popular industry cluster strategies, to the degree they are able to leverage increasing returns and externalities in slow-growth areas, may help reduce regional growth disparities as well. The problem with these applications is that, like the growth center strategies of the 1960s and 1970s, they tend to be successful only where a critical mass of economic activity already exists. Developers might also focus on improving the nonprice competitiveness of local producers (manufacturers and nonmanufacturers) through technical assistance, technology diffusion, training, and so on. Endogenous growth theory suggests a role for economic developers in this regard, but as a means of leveraging externalities and spillovers. With its focus on knowledge spillovers and human capital, endogenous growth theory highlights the critical importance of quality educational institutions from primary and secondary schools through the level of research universities.

III. Elaboration

The traditional one-sector regional growth model has been subject to considerable scrutiny, most of it focused on four general issues embodied in the model's simplistic assumptions: (1) that the interregional reallocation of capital and labor depends solely on factor price differentials; (2) that it is wage rigidity, rather than demand deficiency, that explains persistent regional unemployment; (3) that technological progress (product and process innovations) is instantaneously available everywhere; and (4) that the assumption of constant returns is a reasonable approximation of actual industrial conditions for theory-building purposes. A more detailed review of each criticism is useful for better understanding the nuances of the original model, revisions in the form of multisector models and theories of endogenous growth, and the ways in which the post-Keynesian perspective differs from the neoclassical one. For example, a consideration of the assumption of wage rigidity helps clarify some of the fundamental differences between the neoclassical and post-Keynesian views of persistent unemployment problems, whereas the issue of constant versus increasing returns demonstrates how the emphasis on formal modeling in neoclassical economics can lead to the neglect of some important issues. Note that, as the basic but highly flexible neoclassical model is modified, its implications regarding the growth process seriously challenge the prediction of regional convergence in growth rates and the associated call for minimal government intervention.

Interregional Factor Reallocation

Increases in productivity are possible in the neoclassical model if an initial state of disequilibrium is assumed. If it is the case, for some reason, that factors are allocated inefficiently (reflected in regional differences in rates of return), a possible source of such productivity increases is the reallocation of factors between regions (McCombie 1988a). The neoclassical model predicts that labor will migrate to high-wage regions, while capital will move where it can obtain the highest rate of return. Thus, a testable implication is that capital, wages, and the capital–labor ratio will grow fastest in low-wage regions, implying that the equilibrating tendencies predicted by the model may actually be at work. A seminal test of these hypotheses by Borts and Stein (1964) led to rejection of the validity of the simple one-sector model on the basis of compelling evidence that the fastest growth of capital and the capital–labor ratio has occurred in *high-wage* regions.[7]

The explanation for the poor performance of the model may lie in the assumption of a single sector. Consider Armstrong and Taylor's (1985, pp. 59-63) simple, two-sector model of regional growth. There are two sectors: a domestic sector producing for local consumption and an export sector responsive to external demand. An increase in demand for the region's exports will raise the price of those exports and, thereby, the marginal revenue products of capital and labor. This, in turn, generates interregional and intersectoral differentials in wages and returns. Capital and labor migration *into* the region, as well as the intersectoral reallocation of labor and capital, will eventually reduce these differentials. Meanwhile, increases in regional income also stimulate demand for the region's domestic product. It should be clear that regions exporting products for which demand exogenously increases will experience net inflows of labor and capital, with subsequent effects on regional growth.

Also questionable is neoclassical theory's exclusive focus on differential factor prices as the determinant of migration. The neglect of impediments to both capital and labor migration, as well as other possible determinants, must be viewed as simplistic: that migration is cost-free, that other barriers to migration do not exist, that labor and capital each possess perfect information regarding factor returns in alternative locations, and that factor prices are perfectly flexible. All are unlikely at best. Migrating workers incur both pecuniary and psychic costs from the transport and/or sale of assets (for example, their homes). Although investment capital can be safely regarded as highly mobile, existing firms (with their associated plant, equipment, and trained labor force) cannot be (Armstrong and Taylor 1978). Even in an age of advanced information technology, the assumption of perfect information is unrealistic. Moreover, firms would incur tremendous costs (as would workers) if they attempted to calculate factor returns in every possible location.

Nevertheless, the one-sector model is not entirely without empirical support. The model is reasonably successful in explaining the variation in growth rates across states; capital and labor are shown to respond more to differential factor returns than to employment opportunities, although the latter is still a statistically significant determinant.[8] Dynamic simulations of the model indicate a tendency toward convergence of regional growth rates of output and productivity, as predicted by the theory. It is noteworthy that a demand-oriented model has been proven to have more explanatory power in a test using the same data set.[9] Moreover, researchers applying growth theory to Indonesian regions found that the influence of quantity effects (the employment opportunity specification) is such that regional growth *diverges* in simulations over time (Giarratani and Soeroso 1985). This is because the growth-induced employment opportunities are stronger in the growing regions of Indonesia, creating a positive feedback mechanism that overpowers the equilibrating tendencies of factor flows assumed in neoclassical theory.[10]

Regional Unemployment

Another source of sustained criticism of the neoclassical regional growth model, and of neoclassical economics in general, is the neglect of possible demand deficiencies as a cause of unemployment. Because the neoclassical model assumes that factor prices are sufficiently flexible that capital and labor remain fully employed, involuntary regional unemployment is technically impossible. This approach is not particularly helpful in the regional context since persistent unemployment problems are frequently the primary concern of economic development practice (Richardson 1978). In the neoclassical world, regional unemployment can only be a result of wage rigidities, perhaps caused or exacerbated by collective bargaining agreements, minimum wage legislation, and high unemployment benefits (McCombie 1988b). The implication is that, to lower unemployment, wages must be reduced, either through subsidies to firms, limitations on trade union activity, or the elimination of benefits.

McCombie (1988b) provides a useful analysis of the relationship between unemployment and real wages from both the neoclassical and the post-Keynesian perspective. Consider a cut in the real wage in order to restore a full employment equilibrium (the neoclassical prescription in a condition of high unemployment). Firms presumably employ more workers to produce additional output; labor demand and supply are brought back into balance (at a lower level of total employment).[11] Instead, imagine that the wage cut induces a reduction in consumption (after all, workers have less money with which to purchase goods), and therefore demand. Firms reduce production below capacity, laying off workers in the process. Multiplier effects come into play, further reducing production (and employment). In this scenario, it is possible that wages will move pro-cyclically and that those regions with the lowest unemployment rates will offer the highest wages. The actual outcome also depends on interregional effects, since exports may offset internal contractions in demand and wage cuts may influence migration (as well as wage-setting policies in other regions). In any case, the demand-side focus of the post-Keynesian approach should be evident.

Technology Diffusion

Related to the cumulative effects of economic growth in faster-growth regions is the notion that technological progress does not immediately diffuse across space. Technological advances would be expected to be made in more prosperous regions that possess the necessary agglomeration economies to permit significant investments in research and development. Thus, technological progress could enhance the competitiveness of faster-growth regions vis-à-vis lagging or underdeveloped areas. In any case, all regions would not be expected to have equal access to the same technology, due both to the friction of distance and to institutional restrictions (such as patent laws and industrial secrecy). Clearly, the neoclassical assumption that technology is equally available everywhere is a tenuous one. That innovations are not adopted immediately (and tend to diffuse through distinct routes, such as through the urban hierarchy or across buyer–supplier networks), has been partially confirmed by empirical research.[12]

It should be noted, however, that the more integrated the national space economy, the less important delayed technology diffusion should be as a source of differential regional growth. Technology diffusion could be viewed as a case of market failure in the sense that regions do not possess perfect information on all available production techniques (McCombie 1988a). As market integration continues and information flows improve, backward areas should catch up to faster-growth areas; the market mechanism should still lead to an equilibrium with convergence of regional growth rates. The neoclassical model proves flexible enough to account for technology diffusion (in a limited sense). What remains in dispute is whether market integration will occur to the degree necessary to outweigh the positive feedback mechanism that benefits innovating regions while leaving lagging regions behind. If technology diffusion is indeed an important source of regional growth, the implications for economic developers are unclear. Developers might attempt to improve the availability of information regarding new innovations to local firms, as well as encourage research and development activities at home (applicable strategies might include industrial extension, modernization programs, and the establishment of university-business linkages). Research suggests, however, that firms do not necessarily adopt new practices as they become available—even if they know about them.[13]

Increasing Returns and Methodology

As outlined above, a critical difference between traditional neoclassical growth theory and both endogenous growth theory and post-Keynesian theories is the assumption of return to scale. The assertion in early neoclassical models of constant returns in production, which arguably bears little relation to the real world, is sometimes used as a reason to discount entirely such theories as simplistic, naive, and invalid. After all, simple observation would suggest that there are benefits to size; firms clearly seek efficiencies through scale, evidenced by the popularity of expansions, mergers, and acquisitions in industry.[14] Moreover, the returns-to-scale assumption is a critical driver of the results of the model, particularly from a spatial or regional perspective. Constant returns in the basic model suggest a gradual lessening of regional income disparities; increasing returns and associated externalities in post-Keynesian and endogenous growth models can imply polarized regional development, with some communities developing rapidly and others falling behind. The different implications for regional policy—particularly depending on one's view of distributional equity and the importance of place—are significant. But it would be incorrect (and much less instructive) to interpret the early treatment of returns to scale in regional analysis as a story of naive but technically skilled neoclassical economists pitted against gritty and realistic post-Keynesians and cumulative causation theorists. Instead, it is a story of how differences in methodological approach drive model-building and the advancement in our understanding of the dynamics of regional growth and change.

Krugman (1997) makes this point clearly in his brief study of spatial analysis in mainstream economics.[15] Until recently, economists lacked the technical tools needed to model increasing returns in ways that were both tractable and insightful. This problem affected the neoclassical coverage of geographical questions in general; indeed, neoclassical economists focused on the questions they could address within the corpus (and with the mathematical tools) of microeconomic theory, neglecting location theory not because it was regarded as unimportant, but because it could not be handled in a satisfactory way under the prevailing methodology. For similar reasons, "high development theory" (or the study of international development) also stagnated after its heyday in the 1940s and early 1950s. As Krugman notes:

> the basic problem was neither one of ignorance nor of bias. Economists did not abandon the insights of development economics because they had forgotten about the subject; they did not ignore the ideas of geographers because to acknowledge space would somehow conflict with free market prejudices. No, these fields were left untilled because the terrain was seen as unsuitable for the tools at hand. (p. 67)

Endogenous growth models are possible because of improvements in those tools; increasing returns is now a serious subject of neoclassical growth analysis simply because it is now possible to model them rigorously. As for the mainstream economist, the benefits of formal reasoning through expression in mathematics, such that the precepts and implications of theory can be presented in stark relief, far outweigh the costs of neglecting phenomena that fall outside the purview of existing methodological techniques.

It remains to be seen what will be the impact of new endogenous growth theories for understanding regional growth and change. The appearance of such models in regional analysis did not occur until the mid-1990s. But, with their development, as well as continued formalization of post-Keynesian ideas, regional growth theory in general will continue to improve in realism and the capacity to generate policy insights.

Discussion Questions

1. Growth theories often treat capital and labor as "homogenous," or uniform, in their inherent qualities. For example, the models postulate a labor force made up of identical workers. In what ways might

this assumption affect the results of neoclassical and post-Keynesian growth theories?

2. Which is likely to be more mobile, capital or labor? Are there types of capital or labor that are more mobile (able to move between regions) than others?
3. In what ways might changes in telecommunications technologies affect the diffusion of innovations over space? According to neoclassical growth theory, will the economic prospects of peripheral or rural areas improve or decline with improvements in telecommunications technologies? What factors might determine the outcome? Contrast early neoclassical models with endogenous growth theory.
4. The chapter argues that post-Keynesian growth theories are concerned primarily with the adequacy of effective demand. But, in what ways (or what conditions) do the theories substantiate supply-side–oriented economic development strategies?
5. Is there any practicality to government stimulating an exogenous decrease in wages to promote development or address persistent unemployment problems? How might such a decrease be implemented?
6. If one region attempted to somehow reduce wages in a given sector to spur development, would others follow suit? What are the implications of such a process?
7. How might concepts of sustainable development be incorporated into neoclassical and post-Keynesian growth models? Hint: Why are environmental impacts of development often ignored by firms?
8. Regional development theorists traditionally have been concerned with reducing interregional disparities in income by promoting growth in lagging regions. What is the rationale for such a concern? What are the costs and benefits of encouraging labor mobility from depressed to growing regions?
9. Assume that a region with a traditionally strong economy is hit by an economic shock in the form of several major and concurrent plant closures that result in a considerable number of layoffs. According to traditional neoclassical growth theories, what types of adjustments are necessary for the system to return to a low-unemployment equilibrium?
10. What kinds of costs, if any, might be associated with the economic adjustment process postulated in the previous question? Are any costs likely to be short-lived in nature? Are there any that might have longer-lasting effects on the future growth trajectory of the region?
11. The neoclassical perspective suggests that economic developers should be concerned with the efficient allocation of productive resources in their communities. What role should economic developers

take in influencing the distribution of resources—for example, the more equitable distribution of income?

Appendix 6.1

NEOCLASSICAL ECONOMIC THOUGHT

The origins of neoclassical economics can be traced to the development of the concepts of marginal utility and product and the subsequent focus on price determination in competitive markets (Blaug 1968). More generally, the reformulation of classical economic thought during the nineteenth century brought utilitarian ethics and mathematics to the treatment of economic relations. With the introduction of marginal analysis as a means of explaining the allocation of given quantities among competing uses, the emphasis thus turned from long-term growth to near-term equilibrium. Mathematical formalism was taken up most seriously by the Lausanne school, with Leon Walras and his statement of general economic equilibrium as a system of simultaneous equations the best-known contributor. (The so-called Walrasian auctioneer, as a mechanism to achieve equilibrium in all product and factor markets, is discussed in chapter 9.) Carl Menger and the Austrians contributed to marginal analysis with their subjective approach to utility and value, providing one counterpoint to the socialist thinking of the nineteenth and early twentieth centuries by presenting arguments against economic planning and powerful central governments. The seminal work was written by Friedrich von Hayek, who also wrote an influential treatise against central control and planning. The most lasting Austrian influence on regional development is through entrepreneurship theories, most notably, Schumpeter's.

It was Alfred Marshall and his followers (A. C. Pigou, Edward Chamberlin, and J. R. Hicks, for example) who reformulated classical economics most thoroughly and who developed the topics now covered in standard microeconomics texts. Of critical importance was the development of a method (comparative statistics) for analyzing short-run and long-run equilibria. The theory of consumer demand, the theory of the firm, welfare economics, imperfect competition, and the influence of external economies were essential features of this developing neoclassical thought. Given a well-defined set of postulates and assumptions regarding economic behavior, early neoclassical economists aimed to show how the unimpeded operation of the market leads to the maximization of aggregate social welfare. This would be achieved through the market determination of a set of prices that would yield complete market-clearing throughout the economy (called "general equilibrium"). The neoclassical postulates and assumptions include, among others, profit maximization by firms in completely competitive markets, utility maximization by consumers, the notions of diminishing marginal utility and product, and perfectly flexible product and factor prices. The concept of place, or regional economies within a national economic system, remains largely absent from mainstream neoclassical models (Blaug 1997), although this is changing somewhat with recent developments in the analysis of imperfect competition, increasing returns, and growth and trade (Krugman 1997).

Keynes provided the major challenge to the neoclassical perspective; he and his followers formulated a system of thought based on economic aggregates (for example, income and savings) that eventually became macroeconomics (which now includes both Keynesian and non-Keynesian models). His theory discredited the long-standing acceptance of Say's Law and provided a rationale for government intervention to influence aggregate demand and the money supply.

As with any intellectual paradigm, the neoclassical approach embodies certain values and beliefs (or assumed truths) regarding economic behavior and the economic system. Hunt (1989) presents three key values and beliefs that are frequently challenged by other approaches to the study of development. First, the neoclassical approach implies that economic inequality is a major source of incentive; the notion that inequity can impose excessive costs on the economic system is not considered a serious proposition, although adherents to the neoclassical perspective have been concerned with studying the narrowing of inequality. Belief in the incentive effects of inequality does not imply that neoclassical economists embrace the notion that severe inequity is an ideal state of affairs. Second, the neoclassical perspective attaches a high value to individual freedom. Efforts to reduce inequality through infringements on personal freedom are rejected. Third, neoclassical economists believe that the market is a much more efficient allocator of resources than the public sector, even though this commitment to the notion of laissez-faire has not necessarily been vindicated by objective research (Toye 1987).

The field of welfare economics is an important analytical perspective that relies on neoclassical techniques and principles to study the most effective means of achieving policy objectives; therefore, it recognizes the limited role of government in certain circumstances. Because this approach more readily challenges the limitations of the market, the importance attached above to the concept of laissez-faire should, to some degree, be qualified. Hunt (1989) suggests that this approach is not neoclassical, that it actually represents an entirely new paradigm. Again, this is a matter of slight differences in the acceptance of assumptions and postulates; the economic developer need only recognize that the neoclassical approach, like other approaches to understanding local economic change, embodies values that have been, and must continue to be, subject to careful scrutiny.

Economists working within the neoclassical tradition continue to make significant contributions to the analysis of underdeveloped economies, much of it concerned with short-run questions of allocative efficiency. Many of these contributions include the notion of trade on the basis of comparative advantage (see chapter 7), policy impact analysis, and cost-benefit analysis. The traditional neoclassical analysis of long-run economic growth described in this chapter, although not expressly concerned with underdevelopment, has been extremely influential in the United States as a model of subnational development. Indeed, writing in 1973, regional economist Harry Richardson (1973, p. 22) asserted that "neoclassical models have dominated regional growth theory much as they have dominated growth theory in general." Although neoclassical growth theory ceased to dominate regional growth analysis in the 1980s and 1990s (particularly as the development of the theory stagnated within mainstream economics itself), it nevertheless contributed insights and generated some research. Now, with the advent of endogenous growth models, growth theory in the neoclassical tradition has again become one of the most important and promising areas of research and study in regional development (see Stough 1998, Bal and Nijkamp 1998, Krugman 1997).

Appendix 6.2

THE SIMPLE NEOCLASSICAL GROWTH MODEL[16]

An economy in which all variables are changing at a constant proportional rate (for example, 2 percent per year—neither increasing nor decreasing) is experiencing *steady-state* growth. An equilibrium is *stable* if, once the system is disturbed through some type of economic shock (for example, the oil crisis of the early 1970s), natural forces tend to bring the system back into equilibrium. The notion of stable growth requires a balance of two opposing feedback mechanisms. Positive feedback means that the system diverges from some initial state. Growth is usually a positive-feedback process. Negative feedback means that the system regains equilibrium after an equilibrium state is disturbed. Stable growth combines the feedback mechanisms. Positive feedback (growth) moves the system further and further from its initial state, but negative feedback keeps the system on track.[17]

Neoclassical models predicting steady-state growth would be in general accord with observed economic phenomena. Data on the historical growth experience of the major industrial economies show that key economic variables, such as the capital stock, labor, and output, tend to grow at constant, proportional rates. Thus, steady-state growth is an important empirical phenomenon, along with other secondary phenomena, that growth theorists try to explain.[18] Led by Robert Solow (1956), the original developers of the theory were also interested in carefully examining the results of an earlier class of growth models that suggested that steady-state, full-employment growth would probably *not* occur as a normal state of affairs.[19] This conclusion challenged the neoclassical perspective's emphasis on the market mechanism as a certain route to full-employment growth.

Following are the specific assumptions and postulates of the simple one-sector neoclassical growth model:[20]

(a) There is a single good (thus an economy with a single sector, say, manufacturing) whose production may be characterized by a linear function that relates output to factor input supplies (capital and labor, for example) and technology:

$$Y = A(t)\, F(K_t, L_t) \qquad (6.1)$$

where Y, K, and L denote output, capital, and labor, respectively, and subscript t indexes time. $A(t)$ represents technological change. In what follows, we eliminate the time subscript in order to simplify the notation.

(b) The production function 6.1 is characterized by constant returns to scale; that is, a doubling of inputs leads to an exact doubling of output.

(c) Technology and labor grow at the exogenous, constant proportional rates of δ and n, respectively. This puts both variables outside the scope of the model. For our purposes, technology growth can be viewed as the general advance of knowledge. At the national level, ignoring immigration and emigration, labor force growth is determined by the natural increase of population.

(d) What is saved is, by definition, invested; therefore, no separate investment function is needed. This is expressed by the following identity:

$$S \equiv I \qquad (6.2)$$

where S and I denote savings and investment, respectively.

(e) Savings is a constant (exogenously determined) proportion of output:

$$S = sY, \quad 0<s<1 \qquad (6.3)$$

where s is referred to as the average propensity to save, the share of output in a given period that constitutes savings. The growth of capital is simply the increment in new capital acquired by firms due to investment ($I = S$) divided by the existing level of capital (K):

$$\frac{\text{Change in Capital}}{\text{Initial Level of Capital}} = \frac{I}{K} = \dot{K} \tag{6.4}$$

A dot over any variable is a simplified way of denoting a growth rate; thus $\dot{K}$ is the growth rate of capital.

Manipulating the model outlined above yields an expression denoting the growth rate of output as a function of growth in factor supplies and exogenously determined technology (see below):

$$\dot{Y} = \delta + \alpha \dot{K} + (1 - \alpha)n \tag{6.5}$$

where, as noted, dots above $\dot{Y}$ and $\dot{K}$ denote the *growth rates* of output and capital; n is the rate at which the labor force is assumed to grow, α and $(1-\alpha)$ are the elasticities of output with respect to capital and labor, respectively, and δ is the exogenous rate of technological change. Equation 6.5 simply states that the rate of growth in output is due to increases in the growth rates of the labor force, capital stock, and technological change.

A variable of critical interest is labor productivity, or output per worker. Labor productivity may be expressed as $(\dot{Y} - n)$, or the growth in output, less the growth in the labor force. To identify what determines the growth in output per worker, we can rewrite equation 6.5 in the following terms:

$$\dot{Y} = \delta + \alpha \dot{K} + n - \alpha n \tag{6.6}$$

$$\dot{Y} - n = \delta + \alpha(\dot{K} - n) \tag{6.7}$$

Equation 6.7 indicates that labor productivity growth occurs either through technological change, δ, which by definition is determined outside of the model, or through the growth of the capital stock *in excess of* the rate of expansion of the labor force. When capital stock grows faster than the labor force, the ratio of capital to labor increases (more machinery per worker, for example). This process is referred to as "capital deepening."

Up to this point, we have explored the implications of the simple model as stated above. As noted, early neoclassical theorists were primarily interested in the existence and stability of a steady-state growth equilibrium whereby national output grows at a constant, proportional rate. In fact, this neoclassical model describes an economic system that tends toward equilibrium steady-state growth. Although the demonstration of this point is complicated, the important point is that the model implies that output growth in a state of equilibrium is determined solely by technological change and the rate of growth of the labor force. This is expressed:

$$\dot{Y} = \dot{K} = \frac{\delta}{1-\alpha} + n \tag{6.8}$$

The model is stable in that any shock that drives the economy away from the equilibrium will only be temporary; the economy will return to the steady-state equilibrium. In addition, moving n to the left side of equation 6.8 shows that, in steady-state growth, increases in productivity (output per worker) are determined by technological change alone:

$$\dot{Y} - n = \frac{\delta}{(1-\alpha)} \tag{6.9}$$

The national model must be specified in greater detail in order to render it useful for analyzing regional economic change. In a system of cities or regions, increases in regional capital and labor supplies are a function of indigenous local growth as well as migration between regions. Saving rates might vary across regions, as is probably the case, or one might argue that the rates are similar enough for us to assume that all regions effectively save an identical proportion of income in each period. In addition, production technologies and the rate of new technology utilization may or may not be assumed to be identical across regions. Finally, it is possible to deviate (both at the national and regional levels of analysis) from the assumptions of single good and constant returns to scale production functions. Indeed, in the case of regional growth, the two-sector model yields important contrasting implications for the relative growth rates across regions, as do models that incorporate the notion of agglomeration economies through production functions that exhibit increasing returns to scale.

Following are the basic assumptions of the simplest, one-sector model of regional growth:

(a) Production in each region may be described by the same constant returns to scale production function.
(b) While savings in each period S_i varies across regions (regions are denoted by the subscript i) the propensity to save ($s = S_i/Y_i$) is identical everywhere.
(c) The rate of technical progress, $\delta_i = \delta$, is the same in all regions. There is no process of differential technology diffusion.
(d) In the national model, capital and labor growth were determined by s/v and n, respectively, where

$$\dot{K} = \frac{I}{K} = \frac{S/Y}{K/Y} = \frac{s}{v}, \qquad v = \frac{K}{Y} \tag{6.10}$$

In the case of regions, there is the additional influence of migration to consider:

$$\dot{K} = \frac{s}{v_i} + \sum_j c_{ji}, \; n_i = l_i + \sum_j m_{ji} \tag{6.11}$$

where c_{ji} is the annual net capital flow from region j to i divided by the capital stock in region i (the rate of change of net capital flow), m_{ji} is the net migration of workers from j to i divided by region i's labor force (the rate of change in labor due to net migration), and l is the rate of growth of the local labor supply.

As noted, capital and labor are determined solely by regional differences in wage rates and the rates of return to capital. The model does not allow for the possible costs capital and labor may incur in the process of migration. Rather, it embodies the typical neoclassical assumption that economic adjustments will occur with no friction of distance. Regional output and productivity growth in equilibrium are analogous to the national case.

Deriving the Model: Additional Details

To see how equation 6.5 is derived, assume a Cobb-Douglas production function as the specific form of the general production function. Then equation 6.1 can be written,

$$Y = A(t)K_t^{\alpha}L_t^{(1-\alpha)} \tag{6.12}$$

where α and $1-\alpha$ represent the contribution of capital and labor to aggregate output. More specifically, they are the output elasticities and relative factor income shares. Technology is represented in a neutral, disembodied fashion: technological change manifests itself as outward shifts in the production function. Because technology and labor are assumed to grow at the exogenous rates n and δ, we can re-express 6.12 as,

$$Y = A_0e^{\delta t}K_t^{\alpha}(L_0e^{nt})^{(1-\alpha)} \tag{6.13}$$

This is simply a more specific form of equation 6.1. To derive the growth rates, convert 6.13 to logarithms:

$$\ln Y_t = \delta_t \ln A_0 + \alpha \ln K_t + (1-\alpha)\ln L_0 + (1-\alpha)nt \tag{6.14}$$

Differentiating with respect to time, and simplifying, yields the growth of output analogous to equation 6.5:

$$\frac{d(\ln Y_t)}{dt} = \delta + \alpha \frac{d(\ln K_t)}{dt} + (1-\alpha)n \qquad (6.15)$$

$$\frac{dY_t}{dt}\frac{1}{Y_t} = \delta + \alpha \frac{dK_t}{dt}\frac{1}{K_t} + (1-\alpha)n \qquad (6.16)$$

$$\dot{Y} = \delta + \alpha\dot{K} + (1-\alpha)n, \text{ where } \dot{Y} = \frac{dY}{dt}\frac{1}{Y} \text{ and } \dot{K} = \frac{dK}{dt}\frac{1}{K}. \qquad (6.17)$$

Equilibrium Growth in the Neoclassical Model

By definition, the economy is said to be in a *steady-state* equilibrium if output is growing at a constant, proportional rate (for example, 3 percent per year). This requires that the rate of growth of capital equal the rate of growth of output ($\dot{K} = \dot{Y}$). To see why, assume that ($\dot{K} \neq \dot{Y}$). In this case the ratio of the stock of capital to the stock of output (*K/Y*) must be changing. We know that investment is equivalent to the proportion of income saved in the one-sector economy (*sY*), which is, in turn, equal to the simple change in the capital stock ($I = \dot{K}K = dK/dt$), since all investment increases the level of physical capital stock. From the savings and investment relationship, $sY = I = \dot{K}K$ so that $sY/K = \dot{K}$. A changing K/Y implies a changing $\dot{K}$, which, in turn, implies a changing $\dot{Y}$, from equation 6.17. $\dot{Y}$ is determined by $\dot{K}$. Thus, it should be clear that $\dot{K} = \dot{Y}$ is a condition required for steady-state growth.

Now substitute $\dot{Y}$ for $\dot{K}$ in 6.17 and rearrange to solve for the equilibrium growth rate:

$$\dot{Y} = \frac{\delta}{(1-\alpha)} + n = \dot{K} \qquad (6.18)$$

The critical issue from the neoclassical perspective is whether the economy *tends to* a steady-state growth path. In fact, the model described above suggests that it does. Return to equation 6.17 and replace $\dot{K}$ with sY/K:

$$\dot{Y} = \delta + \alpha\frac{sY}{K} + (1-\alpha)n \qquad (6.19)$$

Now consider the case where the change in savings sY/K (and, thus, the rate of growth of capital) exceeds the values necessary to ensure steady-state growth. Since s is constant, K/Y must be below the value required for equilibrium growth. This ratio will adjust automatically to return the system to equilibrium because, by equation 6.19, the rate of growth of capital $\dot{K}$ must be greater than the growth rate of output, $\dot{Y}$, in this situation. Therefore, since the capital stock is growing faster than output, the capital–output ratio must rise to the equilibrium level. A similar argument can be made to show how the capital-output ratio will fall to restore the system to equilibrium if full-employment savings and investment are below the amount required in a condition of steady-state growth.

Notes

1. As the generally unrealistic assumptions of the neoclassical model are relaxed, the models begin to better reflect observed empirical trends. Some view neoclassical theory as providing a picture of how regional economies should operate, whereas post-Keynesian models are more consistent with reality (Chisholm 1990).

2. Chisholm (1990) notes that the reputation of neoclassical theory as a supply-side approach has fostered the notion that the theory provides an adequate description of supply-side issues. In fact, neoclassical growth theory neglects some important aspects of supply. This is a criticism that may also be applied to economic base theory; an exclusive focus on demand does not necessarily demonstrate its satisfactory treatment.

3. Price elasticity describes how changes in quantity respond to changes in price. Perfectly inelastic demand means that the quantity of demand is fixed for the range of relevant prices.

4. This is referred to as the nonaccelerating inflation rate of unemployment (NAIRU).

5. Figure 6.1 is adapted from figure 3.3 in Armstrong and Taylor (1985, p. 57).

6. See, for example, Borts and Stein (1964, pp. 125–61).

7. This was true overall for the study period 1919–1957. Borts and Stein did find that the growth of capital was fastest in low-wage areas within the more limited period 1929–1948.

8. Ghali, Akihama, and Fujiwara (1978) estimated a simple neoclassical model of state output growth differentials in the period 1963–1973. They specified capital and labor growth as dependent upon lagged factor returns and employment opportunities, with the latter proxied by regional differentials in output growth. The lagged specification was meant to account for imperfect information flows.

9. Ghali, Akihama, and Fujiwara (1981) develop a recursive model that includes aggregate supply variables, aggregate demand variables, and factor mobility. They test two versions, with supply or demand factors functioning as the adjustment mecha-

nism; interregional factor mobility is the adjustment mechanism for the supply-side formulation, while change in net exports is the demand-side mechanism. Interestingly, the model generates similar growth rates under a twenty-year simulation, regardless of whether the demand- or supply-side mechanism is used. This suggests that growth models that stress either supply or demand factors may be as useful as more comprehensive models in simulating growth or testing the impacts of public policies.

10. Although the attempt by Ghali et al. (1981) to broaden the explanation of factor migration represents an improvement over the simplest neoclassical specification, one could persuasively argue that an entirely different approach is needed. The human capital explanation of labor migration is one such alternative (Greenwood 1975). Indeed, the influences on migration probably are sufficiently complex to warrant careful submodeling rather than the somewhat trivial incorporation into an already simplified growth model.

11. Note that the fall in the wage is assumed to boost the production of goods for which it is further assumed that there is sufficient demand. This is an invocation of Say's Law.

12. See Berry (1972) and Thwaites and Oakey (1985). McCombie (1988a) provides a useful, brief review of the innovation diffusion literature.

13. Salter (1966) showed that, in some cases, rational profit-maximizing firms may not adopt best-practice technology immediately.

14. Flexible production theories highlight the benefits of smaller-scale production and increased outsourcing (see chapter 10). But see Harrison (1994) for an insightful analysis of corporate flexibility, firm size, and regional growth and change.

15. Paul Krugman is a leading mainstream economist (and new trade theorist) who has argued for better treatment of spatial questions in the main body of economic theory. For a recent summary of his work in economic geography, see Martin and Sunley (1996). For a view of the significance of his work for economic development and regional science, see Isserman (1996).

16. This appendix draws on Hamberg's (1971) mathematical analysis of neoclassical growth. See his chapter 2.

17. Jones (1975, pp. 12–42) provides an elementary review of many of the basic growth theory concepts.

18. The economic regularities that growth models attempt to explain are referred to as the stylized facts of growth. See Branson (1989, pp. 564–68).

19. These are the Harrod–Domar models (Harrod 1939, Domar 1946, 1947); Higgins and Savoie (1995, pp. 76–84) provide a review.

20. The model also assumes perfectly flexible factor prices (thus, full employment of capital and labor), factor substitutability, and diminishing returns to capital and labor.

7

Trade Theory

Understanding trade and its benefits and ramifications for market economies is imperative for local development officials. Major trends in the international economic environment since World War II, including falling transportation and communication costs and reduced tariff and nontariff barriers, have led to continued increases in the volume of trade between nation-states. Between 1960 and 1995, the annual real rate of growth in world exports significantly exceeded the rate of growth in world output (6.1 versus 3.8 percent); in 1960, world exports stood at $629 billion, compared with more than $5 trillion today (both figures in 1995 dollars, from the *Economic Report of the President,* 1997). Although the United States currently exports a relatively modest 7–8 percent of its gross national product (GNP), a share considerably below that of some other industrialized countries, the sheer size of its economy means that it accounts for roughly 15 percent of total world trade (exports plus imports; Berry, Conkling, and Ray 1997). What disaggregated geographic trade data are available suggest that the volume of international trade varies considerably by region in the United States. In 1990, exports as a share of output ranged from 40 percent in the state of Washington to 6.5 percent in Arkansas. Imports as a share of output ranged from 21 percent in Michigan and Massachusetts to 12.8 percent in Oregon (Hayward 1995). Although the extent to which the globalization of the U.S. economy is a recent phenomenon can be exaggerated, international trade is clearly an important part of the national and regional economic picture.[1] That trade is increasingly the focus of local economic development efforts is evidenced by the important place export promotion strategies have assumed in the pantheon of state and local economic development strategies (Erickson 1992).

It is not surprising that trade issues have captured the attention of city and state officials. The reduction of trade barriers through bilateral and multilateral agreements, advances in information technology, and the subsequent steady increase in international economic linkages have led some analysts (for example, Porter 1990 and Hayward 1995) to assert that cities and regions—not nation-states—are the relevant geographical economic units on the world stage. In his recent book, *The End of the Nation State*, Kenichi Ohmae argues that the global economy is currently undergoing a transition from an industrial to an information age, such that the need for and effectiveness of mechanisms of centralized control by national governments is gradually being eliminated. Already, through the new systems of communication and information transfer, multinational companies regularly (and legally) subvert attempts to regulate and channel their activities to the benefit of their home countries. As the argument goes, in the information age, the major players will not and should not be nations, but instead dynamic regional economies, many of which span national borders (for example, San Diego–Tijuana and Seattle–Vancouver). According to Ohmae, national governments should abandon traditional redistributive programs and industrial policies that favor specific industries and prop up declining rural areas. In the process, they should "cede meaningful operational autonomy to the wealth-generating region states that lie within or across their borders, to catalyze the efforts of those region states to seek out global solutions, and to harness their distinctive ability to put global logic first and to function as ports of entry to the global economy"(1995, p. 142). While not beyond dispute, Ohmae's ideas highlight the global-local dynamic that characterizes much of the recent work on international trade and its geographic implications (Moss Kanter 1995).

Trade theory constitutes an increasingly diverse body of work that attempts to provide answers to several fundamental questions regarding cross-border economic transactions (Krugman 1990): (1) Why do countries or regions engage in trade? (2) What determines the international (and interregional) pattern of specialization? (3) What are the effects of protectionist measures? and (4) What is the optimal trade policy? Although neoclassical theories of comparative advantage and factor proportions remain at the core of trade theory, much recent work examines trade in the context of increasing returns and imperfect competition. Unlike traditional theory, which makes a strong case for free trade and minimal government interference, the "new trade theory"suggests that, in some cases, a degree of managed trade can generate gains. Indeed, like the new, or endogenous, growth theory described in chapter 6, models of trade in a world of increasing returns can yield results strikingly consistent with those of the cumulative causation theorists. These models suggest that, without some intervention to stem their decline, some nations (regions) may be left permanently behind. Even so, given the complexity of the international economic environment, most economists remain pessimistic about the likely effectiveness of government attempts to influence trade. Paul Krugman describes new trade theorists as "'cautious non-activists'—willing to do research on strategic trade policy, but not to propose actually doing it, at least not right now" (1996, p. 110).

To understand local development, trade theory must be applied both in an interregional and international context. The interregional case, however, is not necessarily synonymous with the international one. Obviously, trade effectively takes place between subnational units (states, counties, and metropolitan areas), as well as between those units and other countries. Trade theory is related to economic base, regional growth, and (as will be shown in the next chapter) product-cycle theory in that it provides a means of understanding regional specialization and the equilibrating (or nonequilibrating) tendencies of factor migration. Yet, key assumptions underlying international trade theory may not hold in the regional case. An example is the degree of mobility of labor and capital, which is typically much stronger between regions of the same country than across national borders. Moreover, state and local governments do not have at their disposal most of the usual policy levers used to manage international trade, including tariffs, quantity restrictions, and currency revaluations. Local development officials must therefore understand trade theory in different contexts in order to form a more complete picture of their economy's position vis-à-vis both the national and the global economy. In this chapter, we outline the basic principles of international trade, including comparative advantage, factor proportions theory, and trade with increasing returns.

I. Overview

The first question trade theory attempts to answer is, Why do countries engage in trade? What do they gain from trade that they would lose if they simply attempted to produce all the commodities they require rather than purchase some from other countries? Perhaps this question presupposes too much. Because countries engage in trade, can we assume that they must naturally benefit from it? Such an assumption ignores the fact that countries consist of consumers and producers who make economic decisions based on their own individual preferences. It is not necessarily true that because some individual firms and consumers benefit from the importation of some products and the export of others, the nation as a whole is better off. Greater aggregate welfare might be achieved without trade (autarky), or at least with some restrictions to free trade. Indeed, arguments in favor of tariffs and other barriers to trade take this position. Given a set of reasonably plausible assumptions, it is the principle of comparative advantage that firmly establishes the basis for, and the benefits associated with, international trade. Comparative advantage was first rigorously outlined by David Ricardo in the early nineteenth century. Factor proportions theory, which emerged in the 1930s, explains both the causes and consequences of comparative advantage. Together these concepts remain the central tenets of international trade theory.

Ricardian Trade Theory

Prior to Ricardo, confusion existed about the basis for trade. Adam Smith argued that countries will engage in trade only when each of the partners has an *absolute advantage* in the production of at least one good. A country has an absolute cost advantage in a particular commodity when it can produce that commodity at lower cost than any other country. It is an implicit model of trade under absolute advantage that is invoked by many (though not all) critics of open trade policies, who pit U.S. companies against competitors overseas that enjoy, say, dramatically lower labor costs. Absolute advantage is often considered synonymous with *competitive advantage*, a general term that is rarely defined with care by its users.

Smith argued that international trade may provide a given trading nation's industries with sufficient demand to fully exploit the efficiencies associated with large size. For a firm to grow and reap internal economies of scale requires consumers to purchase large quantities of goods; international markets can provide these consumers even when national markets are limited. By leading countries to specialize in those goods they make at lowest cost, trade promotes an efficient international division of labor. It is this line of reasoning (though not its basis in absolute advantage) that new trade theorists have taken up with the help of modern analytical tools.

Ricardo's analysis overcame confusion about the question of whether nations would engage in trade if one country has an absolute cost advantage in the production of all tradable goods and services. In outlining his theory, Ricardo made a number of assumptions: (1) that transporting goods between countries is costless; (2) that there are no artificial barriers to trade (such as government quotas or tariffs); (3) that labor is homogenous—of comparable skill per unit; (4) that the market is characterized by perfect competition (there are no increasing returns); (5) that production technologies are identical in each country; and (6) that labor, the only factor of production he considered, is immobile between countries.[2] Employing the classical labor theory of value—that all production costs (or the value of goods) can ultimately be reduced to units of homogenous labor—Ricardo then showed that even if one country has an absolute cost advantage in the production of all goods, it may still engage in trade with other nations based on differences in relative internal economic capabilities. This is a powerful and arguably nonintuitive (at least at face value) finding, because it suggests that a firm in a given country may still find an international market for its goods even if it produces at a higher absolute cost than similar firms overseas. A country has a comparative advantage for the purposes of trade in those commodities which its industry produces most cost effectively relative to other commodities (see Appendix 7.1).

Nations and the regions within them have limited resources (natural resources, workers, capital stock), at least in the short run. In an economy operating at its highest potential rate of output, each good that is produced means some other good cannot be produced. Otherwise, where could the necessary workers be found? There is an opportunity cost associated with specializing in particular goods and industries. If you choose to produce *more* automobiles, it means you have chosen to produce *less* of something else. Given two countries (say, the United States and Mexico) and two goods (computers and sweaters), assume that the United States is able to produce one computer by reducing its production of sweaters by two dozen, while Mexico must reduce sweater output by three dozen just to produce one computer. In this hypothetical case, the United States produces computers at a lower relative cost. However, there is a flip side. Mexico is more efficient in the production of sweaters, since, by making one less computer, it can make three dozen sweaters to the United States's two. The comparative advantage in the United States is computers; in Mexico, sweaters. Both countries benefit if the United States ships computers to Mexico in return for sweaters—that is, if they engage in trade.[3] In fact, in this simple model, the maximum benefits are achieved if the United States specializes entirely in computer production, while Mexico produces only apparel.[4]

Factor Proportions Theory

But why would one country be able to produce a given good at lower relative cost than another country? Ricardo's explanation is simply that labor is less productive in some countries than others. (Remember that labor is the only factor of production in the Ricardian theory.) What, then, determines differences in labor productivity? *Factor proportions theory*, developed initially by Eli Heckscher (1919) and Bertil Ohlin (1933) and further refined by Paul Samuelson (1948), postulates a more sophisticated framework for understanding the causes of comparative advantage. Essentially, factor proportions theory states that countries trade because they "are different" in terms of their endowments of the equipment, materials, personnel, and expertise that go into producing goods (Krugman 1990). Countries that have more land relative to workers (for example, Canada) will find their comparative advantage in land-intensive goods such as wheat. Countries that have more low-skilled workers relative to high-skilled workers (for instance, Mexico) will export lower-technology products (such as apparel). Countries well endowed in capital and skilled workers (the United States and Japan) will find their comparative advantage in such goods as electronics, computers, and laboratory instruments. And so on.

Given a set of restrictive assumptions, factor proportions theory also demonstrates that international trade will tend to equalize factor and commodity prices over space even in the absence of factor migration. Samuelson (1948) provided the seminal exposition of this result, called the *factor price equalization theorem*. Under factor proportions theory, differences in the relative costs of production of goods across countries are due to differences in the relative scarcity (and therefore price) of factors. We can imagine, however, that once trade ensues (on the basis of those differences), factor prices across trading partners will eventually converge. In other words, although factor prices in the absence of trade first establish the basis for trade, they are not likely to remain stable once trade ensues.

Consider the United States and Mexico again, which, in our hypothetical example, engage in the trade of capital-intensive computers and labor-intensive apparel. The United States, which is relatively well endowed with capital, exports computers to Mexico, which has a large population but relatively little capital. Mexico, with its comparative advantage in labor-intensive products, exports apparel to the United States. In order to satisfy U.S. demand for imported apparel, Mexico gradually transfers resources from its own home computer industry. That is, Mexico begins to specialize in the product where it has a comparative advantage. Because computer production is capital-intensive, relatively more capital than labor is released for use in the production of apparel. This stimulates an increase in the price of the labor, while generating a decrease in the price of the scarce factor, capital. It is harder to find workers in Mexico as the apparel industry grows, so each worker can command a higher wage. Yet capital is relatively easier to secure. Although capital is being released from the computers sector, apparel manufacturers need only a limited quantity of it. Therefore, its price falls.

The opposite process occurs in the United States. As resources are shifted from the apparel to the computer industry as the United States specializes in its export good, the price of the relatively scarce factor—labor—decreases. This occurs because labor is being released from apparel production; the subsequent increase in the labor supply brings down the wage in the computer industry (former apparel workers now seek jobs making computers). Note that these processes change the distribution of income among labor and capital. The introduction of trade creates both winners and losers; yet the gains to the winners presumably outweigh the losses to the losers. Factor proportions theory holds that free trade leads to an *overall* increase in welfare.[5] Eventually, if we assume that industries produce with constant returns to scale, the price of capital and the wage rate will converge to equality between the United States and Mexico.

New Trade Theory

Because the assumptions of Heckscher-Ohlin theory are highly restrictive, its predictions are at least open to question. The theory assumes two factors of production—capital and labor—both of which are mobile between industries but immobile between regions. This important interregional immobility assumption means that, while the price of each factor must be identical in each industry within a region, the prices may differ between regions. Factor and commodity markets are characterized by perfect competition, no trade barriers exist, and transport costs are negligible. The technological methods of production (production functions) for each industry are identical across regions so that, when faced with the same factor prices, industries in each region will select the same combination of capital and labor (or identical capital-labor ratios). Production functions are also assumed to exhibit constant returns to scale, which rules out possible agglomeration economies. Finally, computer production is capital-intensive (relative to apparel), while apparel production is labor-intensive (relative to computers), at all possible sets of factor prices. This assumption of *strong factor intensity* is critical to the unambiguous results generated from the model. Capital-intensive commodities might become labor-intensive once the ratio of capital to labor prices becomes very high. This means that some commodities could be produced using a labor-intensive process in labor-rich regions and capital-intensive process in capital-rich regions. In this case, the Heckscher-Ohlin theory could predict no distinct pattern of trade (Armstrong and Taylor 1985).

The new trade theory introduces another explanation why countries (and regions) trade: because there are economies of scale in specialization. For some goods—say satellite launching rockets, large passenger aircraft, ocean liners, deep-sea oil-drilling equipment—economies of scale are so significant that the world market can bear only a few centers of production (Krugman 1990). The theory helps explain why trade is most common between countries that are the most similar in their factor endowments. The theory also places much emphasis on the role of chance in determining the pattern of specialization. Some countries and regions get lucky; by virtual serendipity, production of a particular good is started there first. Then, by virtue of economies of scale (internal to the plants plus external economies of agglomeration), the region extends its advantage vis-à-vis other competitors. In particular regions, production of certain goods may become "locked-in." As should now be clear, new trade theory is closely related to the new endogenous growth theory discussed in chapter 6. Although new trade theorists recognize that it is theoretically possible that trade with external economies may be detrimental to some regions and countries, they argue that few regions would fall behind permanently. As Krugman and Obstfeld note:

> Canada might be better off if Silicon Valley were near Toronto instead of San Francisco; Germany might be better off if the City (London's financial district, which, along with Wall Street, dominates world financial markets) could be moved to Frankfurt. The world as a whole is, however, more efficient and thus richer because international trade allows nations to specialize in different industries and thus reap the gains from external economies as well as the gains from comparative advantage. (1997, p. 152)

II. Applications

Outside the United States, neoclassical trade and growth theory are the mainstays of development planning, especially development promulgated by multilateral aid organizations. Benefit-cost analysis, widely used to evaluate investment projects in developing countries, is based squarely on neoclassical thought.[6] Within the United States, economists use trade theory to advocate less government intervention, in order to promote freer international trade and fewer restrictions on interstate trade. Local economic developers, on the other hand, often ignore the implications of trade theory and end up supporting protectionist measures and growth strategies that do not always improve the economic well-being of the community.

Trade theory narrows the economic developer's attention to tradeable commodities, both goods and services; the large group of commodities either not exchanged or minimally traded may be ignored. For traded commodities, the developer may try to determine whether absolute advantage or comparative advantage exists by comparing the relative prices of traded commodities to world prices or to prices in those places with which the region trades.[7] Usually, however, the number of commodities traded will be too numerous to examine individually. For simplicity, the developer may want to assume that local comparative advantage exists for internationally traded goods and that absolute advantage exists for interregionally traded goods.[8]

Because trade theory is complex and operates at the commodity level, developers should consider a shortcut. They can assume that existing regional specializations reflect absolute or comparative advantage. With this assumption, they could challenge policies that threaten regional specializations because these can reduce the region's comparative or absolute advantage. In general, the developer would work toward greater efficiency and productivity in the local economy while at the same time opposing barriers to trade. Exports should be diversified only after carefully assessing the competitive exports of other regions.[9] Similarly, import substitution should be carefully compared to further export-sector specialization. More productivity gains may result when exports are used to purchase imports with foreign exchange earnings than when the imports are produced locally. The theory strongly supports local infrastructure development, improvement in government efficiency, and other measures that could increase the productivity and lower input costs for all local producers. For example, improvements in intraregional transportation can reduce costs to local companies while interregional transportation improvements (or lower freight rates) can expand trade (market) areas for basic-sector companies.

Many metropolitan economies, with thousands of traded commodities to consider, are far too large and complex to examine each traded commodity in turn; working at the two- or three-digit industry level is more practical. The applications of trade theory at this level, on the one hand, lead to the same recommended strategies as were supported by neoclassical growth theory: opposing barriers to trade; promoting greater efficiency and productivity in the economy, including efficient government that would benefit all local producers; and supporting local infrastructure development, especially improvements in intraregional and interregional transportation. This consistency should not be surprising, since both theories are drawn from the same set of basic assumptions.

On the other hand, trade theory forces developers to analyze these strategies in greater detail. The local developer might try to build on existing specializations in manufacturing and selected financial, business, and personal services. But he or she would be able to support these specializations only in ways that promote economic efficiency. For example, the developer might pursue an export expansion strategy, but only by helping exporting companies become more efficient producers. Export expansion led by more efficient local companies probably would generate greater benefits than an import substitution strategy. As we have seen, local consumers are better off buying relatively cheap imports than buying the more expensive local products that replaced these imports.

Most developers face an important strategic decision in applying trade theory: is it better to specialize or to diversify? Much research supports the virtues of each, and local politicians usually seek the benefits of both. The basic tenet of trade theory is that a region (or individual or firm for that matter) can best achieve welfare gains if it specializes and trades; yet more diverse regions tend to experience greater stability and, often, more sustained growth. To deal with the tradeoffs, economic developers should understand that diversity is not the absence of specialization; rather, diversity is the presence of *multiple* specializations.

III. Elaboration

Trade between nations and regions can be described as a sequence of trade-based regional growth: (1) closed economy—self-sufficiency without trade; (2) open economy without factor mobility—the international case; (3) open economy with factor mobility—the interregional case; and (4) regional autarky. Given two self-sufficient regions at pre-trade equilibria, the potential for trade between them exists if equilibrium prices in the two regions are unequal. The different equilibria may reflect unequal factor endowments, unequal tastes, or both. In the short run, commodities are traded until a new equilibrium price is reached. This price consists of the terms of trade; at this price, there is a balance of trade—that is, for both regions exports equal imports. In the long run, specialization toward the commodity in which the region has a comparative advantage further changes the mix of commodities produced, compared to the mix consumed in each region. Specialization will continue until all increases in welfare are realized through trade. Thus, both regions gain by engaging in trade based on comparative advantage.

As described above, under highly restrictive assumptions, free trade can lead to factor price equalization between the two regions. Wages and profit rates may tend toward equality even without factor mobility between the regions (the international case). As a result of commodity-based trade, workers with similar jobs would earn about the same real income in both regions. In the interregional case, however, the two economies are open to the migration of labor and capital. Factor mobility can equalize factor prices more directly than commodity trade. With labor migration and capital flows, wage and profit rate differentials should eventually disappear.[10]

Ultimately, equal factor prices eliminate the potential for trade. The regions cease trading because further welfare gains are not possible. The regions end where they began—in a state of regional autarky. But, after trade, the level of development in the regions is much more similar. Interregional growth and trade theory, then, offer a strong rationale for open economies. Although the assumptions are restrictive, the regional economy that experiences labor and capital mobility, that specializes in its comparative advantage, and that engages in trade not only increases consumer well-being but impacts factor costs such that returns become more equal compared to other areas. Ideally, per capita income levels in trading regions should converge over time.

Although Heckscher and Ohlin argue that comparative advantage is determined by factor endowments, exactly what constitutes the latter is not clear. Ohlin implies: (1) that factor endowments consist of land (and other natural resources), labor, and capital; (2) that these factors are homogeneous across regions; and (3) that the relative quantities of each determines comparative advantage. Clearly, however, since factors are not homogeneous across regions, meaning that they are not strictly comparable, it may not be possible to ascertain whether a country can compete on the basis of comparative or absolute advantage. In contrast to trade theory assumptions, actual regions have deep cultural, historical, and political differences. At any point in time, regions may experience different levels of development while offering very different, immobile types of infrastructure.

The ambiguous nature of the notion of endowments is evident when one tries to determine when a particular combination of inputs that is unique to a country should determine an absolute, rather than comparative, advantage. For example, it is a source of much concern in the United States whenever labor-intensive production processes depart for areas with cheaper labor. Consider the case of Mauritius, a country that experienced rapid growth in the mid- to late 1980s, as a result of the relocation of textile plants from high-labor-cost areas. Relative labor costs were lower in Mauritius than in many other countries when the textile firms moved in. If one assumes that this is the sole reason the firms relocated, the question is, does the lower cost of labor in Mauritius constitute a comparative or an absolute advantage?

The answer may be found in why labor is more expensive in other countries, such as the United States. The reason could be scarcity of labor, or scarcity of this particular type of low-skilled labor. However, institutional factors such as minimum wages, unemployment insurance, and worker health and safety regulations also play an important role in the price of labor. Moreover, historical and cultural factors may affect what is considered an acceptable wage rate in an area. If the difference in labor price between the United States and Mauritius is due to scarcity, then the production of textiles in that region may be due to comparative advantage. Otherwise Mauritius's competitive position is due to absolute advantage since textile firms are not responding to the availability of resources but, rather, to the environmental factors unique to each country.[11]

Even if a distinction between comparative and absolute advantage makes sense, nations trade on the basis of comparative advantage only if exchange rates can be adjusted to ensure balance-of-payments equilibrium.[12] Typically, regions may not be able to trade on the basis of comparative advantage because, first, they possess no mechanisms that can achieve exchange rate adjustments; and, second, demand or institutional factors may generate prices that do not reflect regional productivity differentials (McCrone 1969). The significance of this problem depends on whether or not regions are likely to experience significant balance-of-payments problems. With the obvious data shortage, empirical work on this issue is limited, while theorists remain divided (Armstrong and Taylor 1978).[13]

According to the neoclassical approach, because factors are perfectly mobile between regions, wage and profit rates should be approximately equal and no unemployment should exist. In this situation, trade should be based only on absolute advantage. However, if one region is more productive than another across all industries, the inefficient region will be unable to establish competitive prices. As a result, production in this region must contract; surplus labor and capital will tend to out-migrate rather than become unemployed:

> Thus there is a contrast between international trade where differences in productivity can be matched by differences in factor earnings, where exchange rates can be adjusted to ensure trade can take place on the basis of comparative advantage, and where all participants in trade no matter what their level of efficiency or their endowments may enjoy economic growth. And, on the other hand, inter-regional trade with complete factor mobility which implies equality of factor earnings, trade only on the basis of absolute advantage, no regional unemployment, but a tendency for regions with below average efficiency to decline while others expand. (McCrone 1969, p. 79)

Declining demand for local products may further the unproductive region's decline via the multiplier, while a negative balance of trade can also encourage a downward spiral. If labor fails to migrate, surplus resident labor may force down wage rates or unemployment may persist. This is a positive feedback system that works as a vicious cycle of interregional divergence and decline similar to Myrdal's (1957) backwash effects. There may be too much factor mobility for comparative advantage-based trade but too little mobility for no unemployment. In short, if a region tends to trade on the basis of absolute advantage, and it has no such advantage in any commodity, its economy will contract.

This raises the question of whether a less-developed region would be better off as a less-developed country. With the status of a nation-state, the region could pursue an independent economic policy more suitable to its needs by setting exchange rates given the balance-of-payments situation, establishing protective tariffs, and implementing monetary and fiscal policy. At the same time, national boundaries would limit the factor movements that act as an automatic economic adjustment mechanism. The new nation would forfeit the benefits of cross-subsidization from other regions although it might gain transfers from international aid organizations (at the likely cost of some sovereignty over economic policy). Moreover, as a nation, the region would have to face world prices; this would require a painful readjustment for any industries accustomed to price subsidies. The point of analyzing the differences between a sovereign region or one within a country is not to advocate secession. The comparison points, instead, to export-oriented growth policies for declining regions. One approach is to assist regions in simulating economic sovereignty and comparative advantage, perhaps through regionally specific taxes or subsidies. One should note, however, that this argument is highly abstract and must be grounded in fact before it can confidently be used to understand what is going on in a specific regional location.

Ultimately, McCrone's (1969) argument concerning the benefit of exchange rates that ensure balance-of-payments equilibrium is not compelling. Income transfers enable a region to have a continually deteriorating balance-of-payments position; if income were not being sent into the region, there would be no "foreign exchange" for obtaining commodities from outside the region. A separate currency might be valuable, however, to the extent that it allowed a region to set its monetary and fiscal policies without worrying about the consequences of external inflation, assuming such policies are an effective way to stimulate growth.

One of the most powerful conclusions in trade theory—that gains from trade are mutually beneficial—is also questionable. Certainly, specialization caused by trade increases aggregate production, but the distribution of these gains may not benefit both trading partners. Let us again consider two countries, the United States and Mexico, that produce two commodities, computers and apparel. Assume that computers are income-elastic, whereas apparel is income-inelastic. Relative price differences lead to trade, with the United States buying apparel from Mexico and the latter importing American computers. In the short run, trade continues until an equilibrium price level is reached and both countries are better off; in the long run, the two countries specialize, expanding aggregate output as described in section I of this chapter.

As output expands, the prices of the two commodities fall, causing real incomes to rise. However, as incomes rise, a greater proportion of income is being spent on computers than on apparel. Although consumers may still consume a greater quantity of apparel, the amount they spend on this commodity may actually fall. Moreover, any further increases in apparel production by Mexico may cause this situation to worsen. Thirlwall (1980) identified this problem with regard to the export of income-inelastic goods. At this point, it might be very difficult for Mexico to shift back to the production of computers if the United States successfully developed its resources to become even more efficient in computer production. Thus, in the long run, the net gains from trade may not be mutually beneficial.

That trade may not be mutually beneficial represents a fundamental challenge to the notion of equilibrium and regional convergence that underlies neoclassical theory. The theory of cumulative causation suggests, in short, that success breeds success, while failure breeds failure. Labor, capital, and trade tend to favor rich regions, to the detriment of poor regions, so that market forces reinforce the concentration of growth in prosperous metropolitan centers and regions. Skilled labor migrates from poor areas; capital seeks locations where demand is highest. Although wealthy regions needing goods from other regions for expansion may diffuse benefits to poorer areas (called "spread effects"), such trade generally will be more beneficial to the more prosperous regions because they are likely to be experiencing external economies and higher productivity that reduce poorer regions' gains from trade. Lagging regions experience price-inelastic demand for their exports, capital inflows to industries that function as economic enclaves, and the suppression of their industrial bases that are unable to compete with imported products. With surplus labor, productivity improvements tend to lower prices, while inelastic demand restricts the expansion of markets. As a result, interregional trade negatively affects poor regions (backwash effects).[14] In effect, cumulative causation theory postulates that economic systems are inherently unstable, due to forces that tend to reinforce regional income differences rather than mitigate them.

One could argue that the diseconomies of excessive growth in prosperous regions—such as congestion, environmental problems, strains on public services, and housing shortages—should at some point outweigh the economies of scale and the agglomeration economies that often accompany growth and lead to slower growth rates. However, because these social costs are not accounted for by individual producers, they are not reflected in prices. Thus, growth may well continue where social costs are greater than the private benefits of increased production, yet bypass lagging regions where underutilized labor and infrastructure are available.[15] Although the state has often functioned to introduce and support exchange and free trade, it may have a legitimate role in modifying and mitigating the negative results of interregional exchange by developing programs to counteract backwash effects and reduce regional inequality.

These criticisms yield two important implications for economic developers seeking to apply trade theory and develop an export base that can help a region benefit from trade. First, comparative advantage is appealing in theory but difficult to determine in practice, and thus it may not be especially useful for guiding regional policy. A region might be better off focusing on the development of industries in which it possesses an absolute advantage in relevant markets. Second, to the degree they can be identified, there may be reasons to discriminate between industries in which a region has a comparative advantage, as opposed to simply promoting all those sectors on the basis of trade theory. In particular, the production of income-elastic commodities for export probably should be encouraged more than those that are income-inelastic.

Discussion Questions

1. What is the difference between comparative and absolute advantage? How is competitive advantage typically defined in the popular press or academic literature? Which of those three concepts are most useful for characterizing international and regional trading patterns of industries in your community?
2. In what ways are cumulative causation models (chapter 5), new growth theory (chapter 6), and new trade theories related?
3. What are possible sources of externalities that lead some regions to specialize in particular goods, and which, in your view, are most plausible?
4. At what geographic level is interregional trade theory most relevant (for example, census region, state, metropolitan area, county)? In what ways, if any, might trade theory help explain the observed spatial distribution of industries within metropolitan areas?
5. Despite the common discussion of the "emerging global economy" in the popular press, U.S. exports remain at 7 to 8 percent of GNP. Is discussion of the growing importance of international trade overstated from the U.S. perspective?
6. Given what you know from this book's brief introduction to trade theory and the volume of U.S. exports, what is the likely relative impact of free-trade agreements such as the North American Free Trade Agreement (NAFTA) on the U.S. economy? Positive or negative? Relatively large or small?
7. A number of cities and states have set up programs designed to promote international exports. Based on the theory of comparative advantage, what role should state or local officials take in encouraging export activity by local producers? What types of assistance (from information provision to financial aid) would be most effective in sustaining or increasing regional employment and income?

8. Given factor proportions theory, how is it that some countries and regions are still able to export goods that utilize factors in which they are comparatively less well-endowed? For example, how are some U.S. apparel producers able to compete with manufacturers in countries with much lower capital-to-labor ratios? (Apparel production is traditionally a labor-intensive process relative to the production of other goods.)
9. How do differing demand conditions in each trading region affect the distribution of the benefits of trade between regions?

Appendix 7.1

"BASICS OF COMPARATIVE ADVANTAGE AREN'T SO HARD TO LEARN"

Manuel F. Ayau, *Wall Street Journal,* October 20, 1983

As a former member of the protectionist lobby in my country who had the opportunity to learn the error of his ways, I am annoyed with myself for having failed for so long to understand why countries are misled into paying the high cost of restricting trade. Many people, including professional economists who should know better, seem oblivious to the implication of the difference between competitive advantage and comparative advantage, despite the fact that David Ricardo explained it about 200 years ago.

Competitive advantage means, of course, that one party can do something at lower cost than another, and it typically is the basis for protectionist arguments. Comparative advantage is the relationship of one competitive advantage to another, or a comparison among competitive advantages. The uneconomic diversion of resources abetted by failure to comprehend this seemingly elusive distinction is enormous. Getting it right should have high priority because policies based upon correct understanding could alone solve much of the problem of underdevelopment in short order. For instance, to illustrate the magnitude of the damage, I once calculated the irrecoverable cost of trade barriers in Guatemala in the mid-1960s. It exceeded the whole government budget, year after year. In the United States, trade restrictions imposed the equivalent of a hidden tax of $1,200 in 1980 on the average

---

American family of four, according to Washington University's Center for the Study of American Business.

What is rare in economic discussion is a simple example that demonstrates quantitatively the benefits derived from free trade. If the reader wants to do a little arithmetic, I offer the following exercise, based first on barter, and then on the use of money. This is a worst-case example; Japonia has a clear competitive advantage over Latinia in producing both radio and television receivers, as follows:

Man-Hours Required for One Unit

	Japonia	**Latinia**
Radio	1	4
TV	4	8

It follows that 48 man-hours of production results in 24 radios and six televisions in Japonia. The same number of man-hours produces six radios and three TVs in Latinia. Adding, we find 30 radios and nine TVs being produced with 96 man-hours of effort.

Suddenly, Japonia and Latinia choose free trade and tear down the barriers they had erected against each other's products. And miraculously, with the same man-hour requirement per unit and the same number of man-hours devoted to production, their combined output can rise to 32 radios and 10 TVs.

This is not really a miracle. It simply is division of labor based on comparative advantage. Under free trade, Latinia is induced to withdraw resources it had devoted to radio production and concentrate entirely on TVs. Now, with 48 man-hours' input, Latinia produces six TVs and zero radios. Japonia is induced to reallocate some resources. It devotes 32 man-hours to radios, where its competitive advantage is greatest, and the remaining 16 hours to TVs, enabling it to turn out 32 radios and four TVs with every 48 man-hours of effort. The world has more product, but are Japonia and Latinia better off individually? To find out, we have to introduce the price system. In doing that, one thing needs emphasis: It isn't prices per se that count, but price relationships. Differences in price relationships are what people act on.

Here is the lineup of prices (we'll use the same prices both before and after free trade):

	Japonia	**Latinia**
Radio	24,000 yen	600 pesos
TV	96,000 yen	1,200 pesos

After free trade, the Japonian retailer can choose a TV at 96,000 yen or 1,200 pesos, corresponding to an exchange ratio of 80:1. He will want to buy pesos whenever he can get them for less than 80 yen apiece. The Latinian retailer can choose a radio at 24,000 yen or 600 pesos, corresponding to a ratio of 40:1. He will be in the market for yen whenever he can get more than 40 for a peso.

The differential in price relationships between TVs and radios in the two countries has created an entrepreneurial opportunity: buying and selling currencies. Price differentials on many products in addition to radios and TVs, and many other factors, including people's expectations concerning the relative economic outlook of the countries involved, play a part in establishing exchange rates. But the Japonian and Latinian radio and TV marketers should be satisfied if the yen/peso rate falls somewhere between 40:1 and 80:1.

We could choose any number, but let us say that the exchange rate becomes 60:1—right in the middle. Before free trade, a Japonian retailer could buy a shipment of 20 radios and five TVs for 960,000 yen. A Latinian retailer could buy the same shipment for 18,000 pesos. After free trade, the Japonian and Latinian retailers, each acting in his own self-interest, do their buying. Here is the result:

Japonian retailer:	
20 radios × 24,000 yen	= 480,000 yen
5 TVs × 1,200 pesos x 60 yen	= 360,000 yen
Shipment:	= 840,000 yen
Saving: 120,000 yen	

Latinian retailer:	
20 radios × 24,000 yen/60 pesos	= 8,000 pesos
5 TVs × 1,200 pesos	= 6,000 pesos
Shipment:	= 14,000 pesos
Saving: 4,000 pesos	

In both countries, purchasing power has been increased. Both can afford to buy more of the same things, or to buy new things they could not afford before. Both are wealthier.

Possibly you aren't convinced until you can see it "in dollars and cents." So why not create a world price, in dollars, for radios and TVs and do the arithmetic over again? At a 60:1 yen/peso exchange rate, the dollar price of a radio is $80 and the dollar price of a TV is $240, based on 300 yen equals five pesos equals $1. You will find that Japonia will have enough extra radios to sell at $80 each to buy from Latinia the TVs it stopped producing, and still have some dollars to spare. And Latinia will have dollars left over after selling extra TVs to buy all of the radios it no longer makes.

This is a severe test because neither radios nor TVs are sold at the "average"price (nor are they produced at the "average"cost). When voluntary exchange is the rule, these products, like all others, are produced at marginal cost and sold at the price the market will pay for the next increment of output. If we allow ourselves to think in terms of averages, it is easy to jump to conclusions.

One such conclusion is that our industry will collapse because it can't compete with their industry. It is true that some radio and TV manufacturers will not be able to compete with the lower-priced foreign products. It is also true that some manufacturers may have to close down some of their more antiquated production facilities. In either event, it is the marginal use of resources that must be relinquished, and that act could add to, rather than subtract from, well-being.

There is a myth, which has about as many lives as a cat, that countries import and export surpluses. "Dumping" of surplus is a frequent complaint. But the fact is that most exported goods have been produced for export as a means of generating foreign exchange, which when in turn sold, produces a greater income in local currency than if the resources employed in export production had been devoted to production for domestic use.

I often ask students during exams: If our country wants more corn, which should we plant—corn or cotton? The answer, of course, is that if with the cotton that we can sell we can buy more corn than we can produce with the same resources, we should plant cotton to have corn. If we stubbornly insist on producing our own corn, we deny ourselves the advantage of division of labor. Denial of that advantage is exactly the cost that trade barriers impose. But the cost is greater than money can measure because division of labor is the basis of civilization.

Somehow, watching the way the world behaves, it seems that it must be easy to forget the addition to well-being that results from the satisfaction of needs at lower costs. Free trade allows pursuit of lowest cost, liberating resources that generate new consumption demands and new investment and jobs. Protective trade barriers amount to self-inflicted punishment, universally practiced. This tragic, divisive misunderstanding of our age probably will be a mystery to future generations. They will wonder how a period of great technological achievement could possibly have been accompanied by such strenuous efforts to obstruct trade and increase poverty.

Notes

1. Krugman writes in one paper: "It is a late 20th-century conceit that we invented the global economy just yesterday" (1996, p. 207); and in another: "one should have some historical perspective with which to counter the silly claims that our current situation is completely unprecedented: the United States is not now and may never be as open to trade as the United Kingdom has been since the reign of Queen Victoria" (1996, p. 120).

2. The labor mobility assumption is critical. If workers (and multifactor models, other factors of production) can freely move wherever they choose, all production would take place in the lowest-cost country. Ricardo assumed labor is mobile within countries but not between them.

3. To illustrate these principles, as well as show more clearly why nations may trade even when one trading partner has an *absolute* (cost) advantage in the production of all goods, consider the following example. According to the table below, Mexico has an absolute advantage in the production of both apparel and computers since it uses 12 fewer labor hours to produce a unit of apparel and 3 fewer labor hours to produce a computer. But Mexico has to give up four units of apparel, whereas the United States must give up only one unit to produce a computer. Mexico can buy computers from the United States cheaper than it can produce them itself; likewise, the United States will fare better if it specializes in computer production and buys apparel where it is cheapest—from Mexico. According to neoclassical trade theory, then, absolute cost advantages are, in essence, irrelevant to the question of whether countries will engage in trade.

	Labor Hours Required per Unit		Opportunity Cost per Unit	
	Apparel	Computers	Apparel	Computers
Mexico	3	12	1/4 units computers	4 units apparel
United States	15	15	1 unit computers	1 unit apparel

4. The result that regions will completely specialize in the production of the product in which they hold a comparative advantage is dependent on the implicit assumption of constant costs in production. An alternative assumption, such as increasing costs, would yield the more realistic case of incomplete specialization (regions producing some of all commodities). The constant-cost assumption is a result of Ricardo's restriction of the analysis to one factor of production. With only one factor, there can be no diminishing returns; labor productivity remains constant as output expands. With increasing costs, regional prices will converge short of complete specialization. Kreinin (1979) provides an accessible discussion of this issue.

5. We will use the example in note 3 to see why this is the case, but first we must introduce another wrinkle. To calculate the total benefits from trade for each country, we need the equilibrium exchange rate for the two nations after trade takes place. The equilibrium rate is determined by the intensity of demand in each region for the traded commodities; it must fall somewhere between the internal rates of exchange in each region. For example, the United States will sell a computer for any price greater than one unit of apparel (otherwise it could produce apparel just as cheaply itself), whereas Mexico will pay U.S. manufacturers no more than four units of apparel for each computer. (Obviously, one sweater will not buy a computer; the relevant unit measure might, for example, be a gross.) Depending on demand conditions, the realized rate of exchange might be established closer to the Mexican internal rate, in which case the United States would benefit the most from trade. Alternatively, it might fall closer to the American rate. With Mexico paying close to one unit of apparel for each imported computer, well under the four units of apparel that it would cost to produce a computer itself, it would garner the bulk of the benefits of trade.

Assume that the world exchange rate is established at two units of apparel per computer. Assume also that Mexico produces 100 total units of apparel, 50 units for home use and 50 units for export purposes. For its part, assume that the United States produces a total of 60 computers, 35 units of which it consumes and 25 units of which it exports to buy apparel. Given free trade, Mexico consumes 50 units of apparel and 25 computers (each unit of exported apparel buys half a computer—or, it takes two units of apparel to buy a computer; 50 exported units of apparel times 0.5 equals 25). By similar reasoning, the United States consumes 50 units of apparel and 35 computers. In the absence of trade, Mexico could produce only 12.5 computers at home with the equivalent amount of labor used to export 50 units of apparel to the United States. Likewise, American manufacturers could produce only 25 units of apparel locally with the same labor that, in a world of trade, may be used to pay for the import of 50 units. Thus, with free trade, both nations' consumption of apparel and computers is higher.

Note that if the rate of exchange falls closer to Mexico's internal exchange rate, it will be because Mexican demand for American computers is relatively more intense than American demand for Mexican apparel. Therefore, although the United States enjoys greater benefits from trade in exchange terms, the benefits are distributed more equitably in utility (or consumer satisfaction) terms. It should now be clear that the distribution of the benefits of trade cannot be analyzed independently of demand considerations.

6. Trade theory is useful in evaluating projects and strategies because it forces comparisons of the advantages of producing a commodity to its opportunity costs. Social benefit–cost analysis is the most appropriate method for making such determinations. By using world prices of tradeables as the shadow prices in the model, the benefits (costs) of local production are compared to the benefits (costs) of trade.

7. Relative prices do not exist in all cases. Commodity exports not available in the receiving region or imports not producible at home are noncompetitive, and, therefore, relative prices do not exist in both regions. The trade of noncompetitive commodities, then, cannot be based on comparative advantage.

8. The *Census of Transportation* is a useful source for determining interregional commodity flows. Regionalized input–output tables can be used (with caution) to distinguish regional trade flows from flows based on national technical relations and to examine the composition of exports and imports.

9. International development agencies have fostered commodity export diversification in less-developed countries (LDCs), only to increase competition and lower relative commodity prices realized by all LDCs.

10. The equality of factor prices reverses the logic of trade theory and leads to equal commodity prices. On the other hand, if transportation costs are assumed to exist, trade can continue without equalizing the terms of trade. Furthermore, increasing returns to scale makes the outcome indeterminate, and government intervention may be required to reach a Pareto optimum.

11. One could also argue that cultural, historical, and institutional factors are part of a region's or country's factor endowment. However, if a certain region is the only one with a particular factor, then trade based on that factor will be due to absolute, rather than comparative, advantage. Moreover, regions are often defined in terms

of some unique factor endowment. This situation makes it nearly impossible ever to find a case in which trade will be based solely on a comparative rather than absolute advantage. See Greytak (1975).

12. Although the principles of trade theory often are illustrated with trade conducted by barter, it is important to understand exchange rates and their relationship to balance of payments. The term *balance of payments* refers to the transactions between a country (or region) and the rest of the world. What the country pays out for purchases of imports must be financed with income from external sources, generally through the sale of exports. Although short-term deficits can be financed by borrowing, persistent balance-of-payments deficits mean that a country (or region) is spending "beyond its means." Eventually, simple accounting dictates that a contraction in expenditures will be required.

We continue the hypothetical example of trade between the United States and Mexico, but this time with consideration given to each trading partner's currency. The left-hand side of the table below provides the production costs in each country, in their own currency, prior to engaging in trade. With these figures, it is possible to determine the limits of the dollar-peso exchange rate. In particular, $1 (equals one U.S. computer) must command 1 to 4 pesos. This range is simply the monetary representation of the barter limits to trade (see notes 3 and 5): it costs American industry one dollar to produce one computer. To be willing to trade with Mexico, the United States must receive at least one unit of apparel in return. Since a unit of apparel in Mexico sells for one peso, the United States must receive at least one peso in exchange for each dollar. Otherwise, it would not have the financing needed to acquire the minimum requirement of apparel. Similar logic may be applied to confirm $1 to 4 pesos as the upper limit of the exchange rate.

The actual exchange rate between the two countries is not established until trade ensues. Consider the three possible exchange rates on the right-hand side of the table: $1 = 2 pesos; $1 = .50 peso, $1 = 5 pesos. Notice that the latter two rates are outside the limits determined by the real cost differentials in the two countries.

			Production Costs in Dollars, 3 Exchange Rates					
	Production Costs/Unit		$1 = 2 Pesos		$1 = .50 Pesos		$1 = 5 Pesos	
	Apparel	Computers	Apparel	Computers	Apparel	Computers	Apparel	Computers
Mexico	1 Peso	4 Pesos	$.50	$2.00	$2.00	$8.00	$.20	$.80
U.S.	$1	$1	$1.00	$1.00	$1.00	$1.00	$1.00	$1.00

Because the first exchange rate, $1 = 2 pesos, falls within the limits determined by real resource cost differences, both countries may preserve their balance-of-payments equilibrium. Mexico is able to finance imports of computers with the money it receives from the export of apparel. A similar situation exists for the United States with regard to the export of computers and import of apparel. Although we have selected an arbitrary rate within the identified limits, the actual rate will be determined by the pattern of demand in the two nations. As long as the rate reflects relative resource costs, balance-of-payments equilibrium will be maintained in each country.

Consider the other two exchange rates. For one reason or another, these rates do not reflect the limits to trade established through differences in resource costs. In the first case ($1 = .50 peso), the peso is (over)valued such that Mexico is uncompetitive in both industries. In the short term, Mexican businesses and consumers will import cheaper American products. The lack of the foreign exchange to pay for them creates an unsustainable balance-of-payments deficit; eventually Mexico will have to revalue its currency to adjust the exchange rate. A similar situation faces the United States when the rate is set at $1 = 5 pesos. Thus, a critical element in the reality of trade between countries is their ability to adjust exchange rates to reflect resource costs. The inability of subnational areas to adjust rates, as members of unified monetary systems with fixed 1:1 exchange rates, is a reason why some researchers have argued that the implications of trade theory have limited relevance for regional economic development.

13. Thirlwall (1980) introduces important basic concepts and considers both trade theory and growth theory from a Keynesian perspective. One concept relates to balance-of-payments adjustments; another is income elasticity of demand.

In contrast to the neoclassical growth model, Thirlwall's model is demand-driven rather than driven by the supply of factors (labor and capital); the balance-of-payments equilibrium may constrain growth due to income adjustments in spite of price adjustments; unemployment is involuntary not voluntary (real wages are too high); and the regional economy may be dominated by positive, not negative, feedback. The latter may hold because of "increasing returns in the broadest sense" (1980, p. 421). This Keynesian model may look like its neoclassical counterparts but it is driven by income effects rather than price effects. Balance-of-payments constrains growth because, unless the region can draw on a permanent source of external financing (transfers), export growth must keep up with import growth.

Thirlwall's model can be used to contrast the logic of economic growth based on price adjustments (price and quantity effects) to the logic based on demand-driven multipliers (income effects). Whereas Keynes considered ex ante investment needed to achieve full-employment saving, Thirlwall examines the ex ante exports needed to equal imports. After showing the Harrod trade multiplier formulation and the tendency for depressed regions to run a balance-of-payments deficit, he notes that:

> countries concerned with regional disparities allow depressed regions to run balance of payments deficits by directing autonomous expenditure and income transfers of various kinds which support consumption and investment in the regions. If the trade sector of a region is large, however, the level of compensation required may be unbearably high to prevent the Harrod trade multiplier from working. (1980, p. 422)

In the model, both exports and imports respond to negative price elasticity and positive income elasticity of demand. The balance-of-payments equilibrium income growth rate (*yb*) is given by the following equation. The key assumptions of competitive markets for traded goods and "the law of one price" reduce the equation to:

$$yb = \frac{e(zt)}{p}$$

where e is the income elasticity of demand for exports, zt is the growth of income outside the region, and p is the income elasticity of demand for imports. Then, using the Verdoorn relationship, Thirlwall relates growth and unemployment. Assuming some inverse relationship between growth and unemployment, he argues that policies should "encourage activities in the region which are as income elastic as possible in markets outside the region" (1980, p. 424). Financial incentives will have positive income effects.

Thirlwall's model supports attention to the export sector—not to realize comparative advantage or input-output linkages, but because export demand is a key part of the regional demand that results in the ability to finance imports. Thirlwall supports export promotion which may proceed like targeted recruitment strategies, only focused on goods with high income elasticities of demand. Thirlwall does not discuss strategies to reduce the income elasticity of imports. Such strategies, which are a function of price effects, may work against the region's short-run comparative advantage, but may build competitive (absolute) advantage in the region. Imports, however, may have to be curtailed to some extent. At least the developer should consider ways to tie imports directly to enhanced export capacity or greater demand for local production.

14. Myrdal (1957) sees spread effects as primarily benefiting the areas near the growth center. Overall, he argues that circular and cumulative causation, rather than balance and equilibrium, will generate increasing differences among regions as the result of the free interplay of market forces. Spread effects are more powerful in rich countries and will lead to interregional convergence, whereas backwash effects are more powerful in poor countries and will maintain or increase interregional inequality.

15. "There is some presumption therefore for supposing that, if left to market processes alone, tendencies to regional concentration of industrial activities will proceed farther than they would have done if 'private costs' were equal to 'social cost' . . . and all economies and diseconomies of production were adequately reflected in the movement of money costs and prices" (Kaldor 1978, p. 149). Consumers may also derive increasing satisfaction from nonmaterial or noncommercial forms of consumption as growth becomes excessive. In this case, a community's level of production could be reduced without decreasing consumer well-being. Negative externalities would decrease as the production of "bads" declined, along with the production of goods. Consumers experiencing less congestion or pollution and more access to natural amenities would enjoy greater satisfaction.

8

Product-Cycle Theory

In the previous two chapters, the theories of economic growth and trade have been presented in detail. Neoclassical growth models and interregional trade theory offer powerful insights about regional economies, but they require fairly restrictive assumptions to retain their predictive generality. The extensions of these theories clearly show how even their basic theoretical conclusions can be reversed in a world of increasing returns or of post-Keynesian wage rigidity.

Although it does not overcome these specific problems, product-cycle theory is useful because it moves us from a world of *given* commodities and factors of production to a world of *changing* production and trade relationships. In product-cycle theory, product development becomes the driving economic force which, like Perroux's growth pole, emphasizes the dynamism of the growth process, but with greater clarity and specificity. Like entrepreneurship theories, the essential dynamic behind product development is innovation.

Product-cycle theory and the entrepreneurship and flexible specialization theories presented in chapters 9 and 10 attempt to model the process of economic *development*. Much of the thinking presented in the previous chapters explains the process of economic *growth*. Theory and practice, often labeled economic development, would be more appropriately called economic growth. Product-cycle, entrepreneurship, and flexible specialization theories help us understand the theoretical and practical differences between growth and development processes. The growth–development distinction is of fundamental importance to economic developers. It is presented in chapter 2 and will be further articulated in chapter 11.

I. The Theory

Raymond Vernon first developed product-cycle theory as a way to explain the "Leontief paradox."[1] In the 1950s, Leontief presented empirical results for the international trade of the United States that contradicted the expected outcomes of trade theory (see chapter 7). The U.S. economy was considered relatively capital-intensive, yet empirical results showed that the United States exported relatively labor-intensive products and imported relatively capital-intensive products. Vernon's product-cycle hypothesis offered one explanation of the paradox. Vernon developed ideas about the product cycle by drawing heavily on his empirical research about the New York metropolitan region.[2]

Product-cycle theory can be explained by contrasting its tenets to basic ideas of interregional trade theory and location theory.[3] Like interregional trade theory, product-cycle theory is grounded in neoclassical economics and considers the interregional pattern of specialization in trade. Unlike trade theory, product-cycle theory is a partial equilibrium argument about the dynamics of development. The theory is more descriptive of actual development dynamics and, concomitantly, less dependent on formal logical argument. It suggests that interregional development patterns are modified over time by recurring cycles of new product, maturing product, and standardized product. Interregional trade does not necessarily lead to convergence of per capita incomes, nor is convergence achieved by introducing labor mobility. Consumption differences, production economies, and communication advantages may continue to favor the more-developed region even with complete factor mobility.

Like traditional location theory, product-cycle theory focuses on microeconomic, firm-level decision making (a partial equilibrium framework). Location theory, however, considers the profit-maximizing locations for existing products, most of which are standardized. In general, the theory ignores the factors important to the development of new products.[4]

Product-cycle theory presents stylized facts about the United States, Western Europe, and less-developed countries. In the international case, countries are distinguished by their different industrial structures, levels of technology, factor costs, and consumer tastes. Capital is mobile and labor is immobile between countries. Although all areas have access to modern science and technology, new products tend to originate in the more-developed country, for two reasons. First, entrepreneurs there have better information about commercial opportunities and innovation possibilities because markets are larger and communications channels are more efficient. Second, entrepreneurs are able to exploit the potential markets for more sophisticated new products that exist in the more-developed country. Greater incentives for new product creation exist in the more-developed country because both incomes and wages are relatively high there. High incomes are correlated with more complex consumer tastes; high wages encourage the production of new capital goods that can substitute for expensive labor.

The innovation process results in new products which initially satisfy local demand in the more-developed country. As it matures, the product becomes standardized and producers are able to export the product to less-developed countries. When the production process is completely standardized, the product can be produced in both areas, and competitive pressures drive producers to seek the lower-cost location. In other words, the price-elasticity of demand increases as the product matures.[5]

This diffusion process benefits the less-developed country because the country experiences more rapid economic growth as the result of external investment and trade in standardized products. The nature and rhythm of its growth, however, are set by firms headquartered in the more-developed country. In essence, diffusion leads to a form of economic dependency in less-developed areas because the tastes satisfied and the technology developed through new products are intended to meet the economic realities of the more-developed country and are therefore not always appropriate for less-developed countries. Clearly, unique tastes and resource constraints influence product innovation. For example, manufacturers in the United States have taken the lead in precision instruments and medical devices; German manufacturers are well known for innovation with plastics, while Japanese manufacturers have developed new electronics products with an emphasis on miniaturization.

The location of production in less-developed countries does not follow the trade-theory logic of comparative advantage. Direct investment embodying advanced technology for standardized production processes establishes new plants and equipment in less-developed countries. The decision to invest is stimulated by the desire to defend the market penetrated earlier by exports of the maturing product. Low wages and insufficient capital accumulation in less-developed countries are advantages to outside investors and increase the regions' attractiveness to them. Capital market segmentation may also work in favor of foreign investors with access to the formal capital market segment. These advantages may compensate for the disadvantages of smaller markets, higher transport costs, and fewer external economies in less-developed countries.

The Regional Context

Vernon's original three-country presentation of international trade can be simplified to consider one more-developed region and one less-developed region within the same country. During the new product phase, all production is consumed in the more-developed region. As the product matures, consumption is initiated in the less-developed region and increases as imports continue. Obviously, production exceeds consumption in the more-developed region during the maturing-product phase. It then becomes profitable to locate production of the standardized product in the less-developed region. During this phase, production may drop to again equal consumption in the more-developed region. Later on, the standardized product may be exported back to the more-developed area. If this occurs, standardized production increases to exceed consumption in the less-developed region.

In product-cycle theory, economic development is defined as the creation of new products and the diffusion of standardized products. Development originates in the more-developed region and is exported to the less-developed region through trade and then investment. Establishing a new industry in the less-developed region creates a progressive force that can help eliminate the barriers to interregional equality. Yet product-cycle theory does not predict convergence of regional incomes; the development process can be convergent or divergent.

The diffusion process could benefit the less-developed region absolutely and lead to convergence in several ways. First, access to mature products expands the range of consumption opportunities in the less-developed region. Second, establishing branch facilities that produce standardized products expands job opportunities. Although wages paid for standardized product manufacturing are low compared to wages in the more-developed region, they should be relatively high compared to alternatives in the less-developed region. Third, investments, loans, or grants from the more-developed region should increase in the less-developed region, initially to support imports of mature products and subsequently to construct facilities for the production of standardized products. Fourth, increasing the number of standardized producers could help diversify the local economy and increase its stability. Fifth, multiplier effects could create opportunities for new enterprises in the less-developed region. However, manufacturers of standardized products often require few local inputs beyond labor and physical infrastructure; most are obliged to purchase goods and services from other divisions of the company. Thus, only local businesses providing goods and services to households may grow.

On the other hand, the diffusion process could result in divergent development. Although favorable terms of trade could evolve for the exports of standardized products, this outcome is neither assured nor expected to be long-lasting, given the high price-elasticity of such products. Even if per capita income increased, the inequality between less-developed and more-developed regions is likely to increase because of the growing dependency of the less-developed region on the more-developed region. Consumers and producers in the less-developed region would exercise less and less control over the local economy, since mature products were designed to respond to consumer tastes in the more-developed region and standardized products were developed with the technology of the more-developed region. Relatively high wages in the standardized product sectors could stagnate unless workers found ways to press demands for improvements in wages and working conditions.[6] Furthermore, the branch-plant economy is clearly owned and controlled by outsiders. Eventually, extensive "foreign" ownership would increase the volume of repatriated profits. In some regions, the branch facilities would form an enclave economy with limited beneficial local effects.

Product-cycle theory supports the need for active government intervention to change regional specializations by encouraging the diffusion of standardized products and the creation of new products. The theory supports implementation of public investments in transportation, communications, and other infrastructure. It calls for the dissemination of current information on scientific, technological, and other relevant developments to entrepreneurs. For government in the less-developed region, it appears particularly useful to ensure sufficient exports, permit the importation of maturing products, and train workers and managers for industries producing standardized products.

In summary, product-cycle theory describes the relationships between innovation, structural change, and economic development outcomes. In its regional application, the theory focuses on the spread effects generated by the diffusion of productive investments from the more-developed innovating region to the less-developed region. Firms in the more-developed region introduce new products in response to the high incomes or high labor costs found there. As new products mature, firms seek export markets and find them in the less-developed region. When products become standardized, they can be more competitively produced in the less-developed region. The innovation process required to create new products has developmental impacts in the more-developed region, whereas the investment needed to establish standardized products generates economic growth in the less-developed region.

In the next section, product-cycle theory is applied to economic development practice. Then, the theory is described more fully and criticized.

II. Applications

Economic developers can gain insights by thinking about the concept of economic diversity in terms of product-cycle theory. Among the industrial sectors exporting goods or services, the developer could try to identify the sectors that generate primarily new, maturing, or standardized products. Although product diversity—defined as a good mix of new, maturing, and standardized products—is not easily achieved, the developer at least should become aware of simple diversification strategies that attend solely to increasing the diversity of the industry mix. An area may become increasingly attractive to a variety of industrial sectors over time, but all these industries may produce standardized products. As a result, the area's vulnerability to external competition may increase as it becomes more diverse in term of its mix of industries.

Developers can also assess the competitiveness of their region using the framework provided by product-cycle theory. Some regions may have sufficient size, diversity, and wealth to serve as locations for new product development. Other regions that possess attractively priced human and natural resources may be competitive due to relatively low costs of production. Still others may fail to be competitive for either new or standardized products. From the product-cycle perspective, these regions are the most hard-pressed to establish a meaningful role in the global economy.

Competitive regions in the United States are more likely to function as new product locations than as locations for standardized production when the factors and intermediate inputs are relatively high quality and high cost. To the extent that standardized products are generated from routine production processes, regions outside the United States with lower production costs are likely to be more attractive standardized product locations.

Columbus, Ohio, is used to illustrate the application of product-cycle theory in U.S. regional economies that possess the capacity to serve as a location for the production of new products. Buffalo, New York, on the other hand, has experienced growth historically from the diffusion of standardized products and is currently working to redefine the basis of its competitiveness.

Application to Columbus, Ohio

The economic developer who wants to apply product-cycle theory in its most literal form must try to identify manufacturing companies that can create new products. If these companies can be found, the developer should survey management to determine how well the Columbus economy supports the process of new product development. Companies unable to move forward with new product ideas because of agglomeration diseconomies or other local problems would provide the most useful insights. The developer may be able to mobilize the resources needed to improve the local physical or business infrastructure in ways that would enhance new product development.

A related approach is to divide important local companies into those headquartered in Columbus and those headquartered elsewhere. Developers should survey companies from both groups to see whether they have different business orientations and thus draw on the regional economy's capacity in different ways. The goal would be to identify the agglomeration economies and diseconomies important to companies in these two groups. As before, the general response is to find ways to support the economies and mitigate the diseconomies.

To carry out these strategies successfully, however, may require substantial effort and resources. The developer would need considerable research capability to find the relevant firms and survey their management. To be politically feasible, business infrastructure would have to bestow benefits on more than one company, and such infrastructure could be difficult to mobilize.

More generically, product-cycle theory emphasizes the possibility of competitiveness arising from research and development (R&D), and, generally, from professional education and technical training. The deep pool of university-based talent may provide opportunities for information exchange and joint R&D ventures with local enterprises. Such linkages can encourage new product development.[7] For example, since the 1950s, ceramic manufacturers and the ceramic engineering department at Ohio State University (OSU) have been mutually supportive; metallurgy research with industrial applications has flourished at Battelle; and other information service companies have grown from research-based activities at OSU or local think tanks.

Developers in Columbus could gain further insights by considering product-cycle theory in terms of its implications for business relationships and linkages among local firms. Just as industrial complexes offer firms external economies that lower production and marketing costs, interfirm linkages can offer support for the processes of innovation and customization that often lead to competitive success through new product development. Unfortunately, the research presents no clear way to select important interfirm linkages or support interfirm networks, especially at the local level.[8]

Columbus developers may want to begin this linkage strategy with a few sectors, such as insurance and banking, whose importance is growing in the regional economy. Beginning with the largest companies, developers could ascertain how production and marketing economies are achieved. Perhaps the successful practices that are internal functions of large companies can be stimulated as external functions for a network of small companies through interfirm alliances and cooperative agreements of various kinds. In addition, the economic developer may be able to help young companies directly find ways to cooperate in order to achieve business advantages. If cooperative arrangements succeeded, considerable goodwill and trust would be generated among these firms, which could lead to other cooperative efforts. Yet, strategies to enhance these external economies are not well defined and may prove to be premature or unwieldy in Columbus.

A more straightforward strategy, one consistent with product-cycle theory, would consider the regional economy as a whole. Columbus developers could try to understand the capacity of the regional economy to support the development of new products. Finding ways to enhance this capacity is worth the effort since it appears to be important to Columbus's economic future. Local politicians usually take a less focused approach, instead sponsoring development strategies that work toward conflicting purposes. Strategies designed to improve the quality of Columbus's development capacity are not likely to decrease production costs in that area. For example, efforts to increase the quality of local education and training may improve workforce quality and preparedness but are unlikely to lower labor costs.

In conclusion, strategies for places such as Columbus, Ohio, which are based on product-cycle theory, should focus on the internal development of the regional economy rather than the diffusion of growth from external sources. New product-oriented strategies emphasize R&D, university-business partnerships, business services development, and other efforts to improve the chances of local product or process innovation. Standardized product-oriented strategies, on the other hand, are entirely consistent with those supported by economic base theory: industrial recruitment, efficient government services, and infrastructure investments that would lower production costs.

Although new strategies oriented toward products are attractive for various reasons, these strategies can also be risky. They may generate results too slowly, given the time frames favored by local public officials. Efforts targeted at assisting particular firms may result in protests from firms that do not receive assistance. The strategies require extensive and intensive research that may be beyond the capacity of local economic developers. Economic developers in Columbus may be well advised to seek researchers at OSU or in state government to become allies in the formulation of these strategies.

Application to Buffalo, New York

Economic developers in Buffalo face a different situation. Historically, Buffalo has been an attractive location for standardized products, especially in metals, chemicals, and transportation manufacturing.[9] Over time, however, its competitiveness in manufacturing has eroded. Currently, the region is experiencing an economic transition. Economic developers should consider ways to increase competitiveness through this transition.

One strategy is a form of targeted recruitment to achieve import substitution. Economic developers could analyze imports to identify mature, high-volume, high-value products. Consistent with product-cycle theory, companies exporting these mature products to Buffalo could be recruited to establish branch plants in Buffalo. Local production would, as a result, substitute for local imports. Yet, Buffalo's internal market may not be of sufficient size, and its cost structure may not be sufficiently attractive, to bring in these branch facilities.

Alternatively, economic developers could identify the agglomeration economies related to the standardized products now declining in Buffalo. Pools of skilled labor, suppliers and business services, physical infrastructure, and local knowledge that can generate external economies may remain. If economies remained, developers may be able to encourage the production of capital goods in these sectors—in essence, supplying the technology needed by currently competitive, standardized product-oriented regions. Possibilities may also exist in industrial machinery, electronic and electrical equipment, or industrial instruments.

This strategy is similar to the economic development strategy articulated by Thompson (1968), who suggests that regions can move up "the learning curve" from routine production to precision production. It is a strategy that has worked in other regional economies. For example, the Greenville-Spartanburg, South Carolina, region has undergone a transition from textile-product production to the production of textile machinery and related capital goods. Economic developers would benefit from contacting colleagues who work in manufacturing-oriented economies to discuss potential transition strategies.[10]

Economic developers in Buffalo should consider other strategies that draw on the region's unique strengths. For example, the aging population may offer opportunities for new products or treatments in the health and medical fields. The city's proximity to Canada may suggest ways to increase local production for cross-border trade. Clearly, new directions are needed. The economic future is not bright for regions that are attractive neither for new products nor for standardized products.

III. Elaboration and Criticisms

Product-cycle theory explains that each stage of a product's life requires a distinct set of locational attributes. The characteristics of regions that encourage new products or standardized products are different, as are their locations in the urban hierarchy. New-product locations offer better jobs and have a higher mix of skills than do standardized-product areas, and they tend to attract a greater number of professional and technical workers. High-tech industries or other industries with significant research and development activities are more prevalent. R&D efforts may result in new product innovation. The first manufacturing facilities for new products will be located within the region because product specifications are being finalized. Entrepreneurial activity in these regions is presumed to be high due to better information and "swift and effective" communication resulting from proximity.[11] Furthermore, new spin-off industries are likely to be created from the core of innovative companies. All this activity increases the linkages within the local economy. Regions with active product innovation tend to become relatively high-cost, larger markets that offer agglomeration economies. These locational attributes of product-cycle theory are summarized in figure 8.1.

Vernon (1965) argues that new-product locations must offer firms flexibility in sources and types of inputs and good communication channels to reach consumers, suppliers, and even competitors; comparative costs, emphasized in location theory, are much less important. Standardized production is attracted to regions that offer lower comparative costs, especially abundant lower-cost labor. Facilities are likely to be located in such regions as long as the product's price-elasticity remains high, production is vertically integrated, product obsolescence is unlikely, and the product's value is high relative to weight. Vernon identifies standardized textile products as a good example of such products.

Hekman (1980) broadens the application of product cycle by looking historically at one industry—cotton textile production. His article is an excellent source for identifying the elements needed for product innovation: entrepreneurs and managers (which may be more effect than cause), product designers, engineers, toolmakers, machine builders, and so forth, in contrast to factors typically cited in location theory: labor, transportation, and electric power. The combination of entrepreneurship, skilled mechanics, innovations in management, and textile machinery production resulted in highly immobile agglomeration economies. As the product became standardized, however, the South offered lower costs to support automation and the substitution of unskilled for skilled labor. Thus, the product cycle is useful in explaining the historical development of textiles and certain other manufacturing sectors.

Erickson and Leinbach (1979) adopt product-cycle theory to explain the diffusion of branch plants to nonmetropolitan areas. The factor inputs required for standardized products make nonmetropolitan areas relatively attractive production locations. As the product becomes standardized, major changes in factor inputs occur. First, the scientific knowledge, engineering

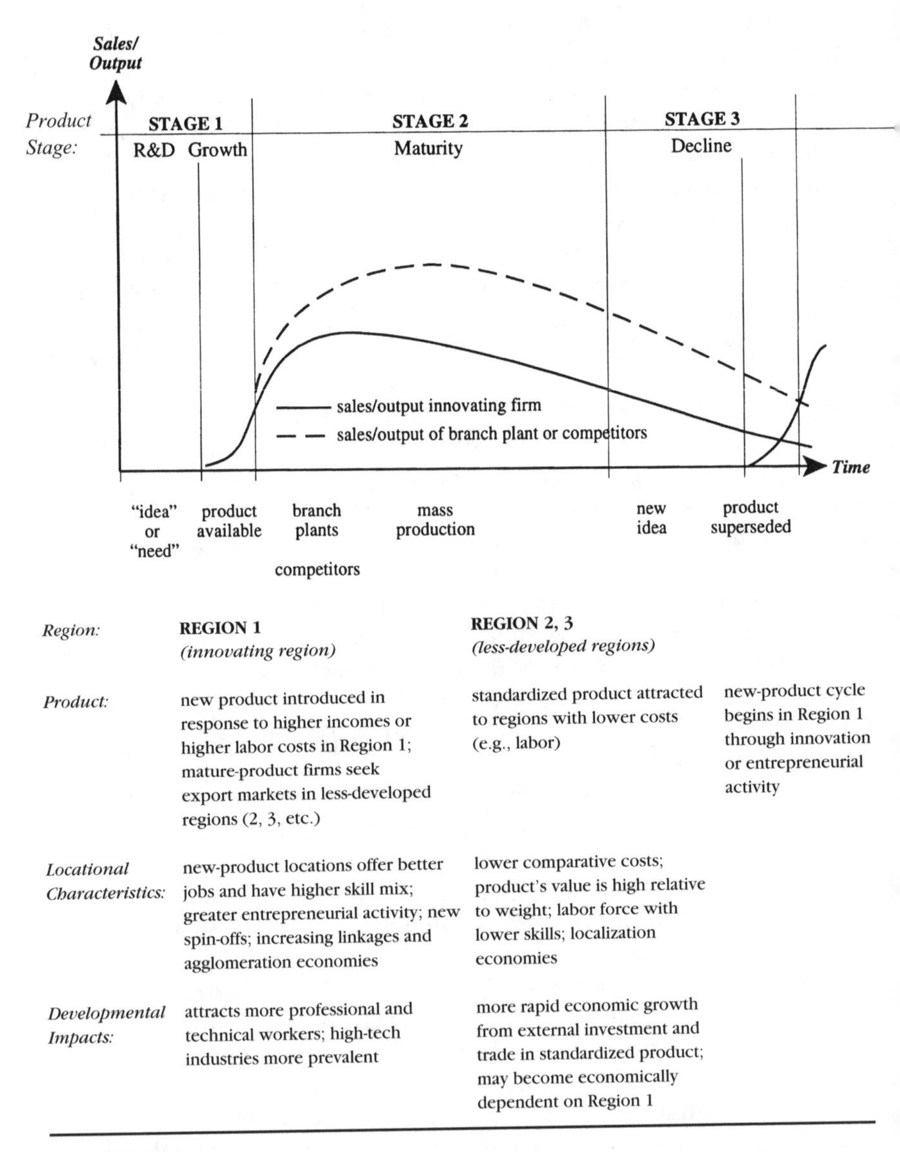

Region:	**REGION 1** *(innovating region)*	**REGION 2, 3** *(less-developed regions)*	
Product:	new product introduced in response to higher incomes or higher labor costs in Region 1; mature-product firms seek export markets in less-developed regions (2, 3, etc.)	standardized product attracted to regions with lower costs (e.g., labor)	new-product cycle begins in Region 1 through innovation or entrepreneurial activity
Locational Characteristics:	new-product locations offer better jobs and have higher skill mix; greater entrepreneurial activity; new spin-offs; increasing linkages and agglomeration economies	lower comparative costs; product's value is high relative to weight; labor force with lower skills; localization economies	
Developmental Impacts:	attracts more professional and technical workers; high-tech industries more prevalent	more rapid economic growth from external investment and trade in standardized product; may become economically dependent on Region 1	

FIGURE 8.1
Product-Cycle Stages and Locational Attributes

applications, and technical expertise required to create the new product decline in importance. Similarly, with standardization, external economies made available to the firm by virtue of its location in a more-developed region increasingly become less important. On the other hand, the demand for production management and unskilled labor grows.

Erickson and Leinbach argue that management inputs are most important to manage the growth process as the new product matures. Management expertise becomes least important for standardized-product manufacturing. Capital is relatively unimportant during the new-product phase, but it becomes important to achieve the growth necessary to export the maturing product and for the facilities needed for standardized-product manufacturing in branch plants. Erickson and Leinbach's argument, however, is unclear about the changes in factor intensity over the product cycle. The new product would appear to be labor-intensive if all types of labor are considered in combination with limited capital inputs. Clearly, the standardized product requires heavy capital investment but also significant amounts of unskilled labor. Capital-intensive production, which Vernon originally posed for standardized products, usually implies high labor productivity and high average wages; production that exploits primarily unskilled labor would usually suggest low labor productivity and wages.

With the publication of Vernon's article, product-cycle theory gained widespread popularity and was adopted in marketing and by economic geographers, who molded it to their particular uses. Additional regional applications are found in Norton and Rees (1979)[12] and Moriarty (1983, 1992).[13] Another group of economic geographers and planners has used the product-cycle model to help them understand the role of technology in regional development.[14]

Criticisms of product-cycle theory have been raised by Vernon (1979) himself, and by Steiner 1987 and Taylor (1986), among others. One general criticism is that, as a partial equilibrium theory, the product-cycle concept may be appropriate only for manufacturing and primary export staples, thereby ignoring the growing service sectors. It is difficult to extend the theory to all exporting sectors, because the product-cycle concept and the reality of export service production are not easily reconciled.

Vernon suggests that the explanatory power of the product-cycle model for international trade patterns has eroded because of the changing international environment. With the rise of the multinational corporation, fewer firms have distinct home markets and related production facilities. U.S.-based firms are more "comfortable" locating outside the home market. Market information on various international markets has become more accessible. Moreover, the large companies that account for the most international investment and trade are often vertically integrated, which lessens the importance of agglomeration economies available in any given location. Furthermore, the previously unique characteristics of the United States—high income and labor scarcity—are no longer unique. Other nations, such as Japan and those in Western Europe, now have economies similar to that of the United States. With the world economies becoming more homogeneous, the potential for the United States to lead in product innovations has declined (Vernon 1979).

Taylor attacks the product-cycle concept on various grounds. First, he considers the theory an example of technological determinism. This criticism is somewhat misplaced. The production of standardized products may be determined by technical, input–output relations, but new and maturing products are shaped more by design, marketing, and management factors than by technology.

Taylor points out that the theory assumes a sufficient market and/or effective marketing, but, in fact, the market may not be able to absorb all production. Furthermore, product differentiation is an important marketing strategy that is not recognized. Neither is product-cycle theory clear on the issue of ownership or control, or even about the definition of the product. It is unrealistic in its treatment of invention and innovation, and, where mass production is the end result, it can show little evidence of products moving through the entire cycle.

Although these criticisms are, to some extent, valid, Taylor and other critics fail to offer an alternative formulation. These criticisms fail to recognize that Vernon never posed product-cycle theory as a general theory of development, or that the theory's simplicity remains one of its main strengths. Rather than trying to integrate concepts relating to the innovation process at the enterprise level, it may be more useful to focus on the definition of the product and product development per se.[15]

Improvisation and New Work

The differences posited in product-cycle theory between new-product regions and standardized-product regions are generally consistent with Jane Jacobs's insights into the economic development process. Jacobs (1969) contrasts efficient Manchester and inefficient but adaptive Birmingham to illustrate the defining characteristics of economic development. In the nineteenth century, Manchester was an industrial powerhouse, with the textile factory of the future. Birmingham was a city of small shopkeepers. Yet, a hundred years later, Manchester was stagnating while Birmingham was thriving. Jacobs argues:

> Manchester's efficient specialization portended stagnation and a profoundly obsolescent city. For the "immensity of its future" proved to consist of immense losses of its markets as other people in other places learned how to spin and weave cotton efficiently too. Manchester developed nothing sufficient to compensate for these lost markets. (1969, p. 88)

Much of Birmingham's success can be traced to its adaptiveness, its ability to generate new work from old work, and to its continuous innovation. Birmingham was precisely the kind of city that seemed to have been outmoded by Manchester:

> Most of Birmingham's manufacturing was carried out by small organizations employing no more than a dozen workmen; many had even fewer. A lot of these little organizations did bits and pieces of work for other little organizations. They were not rationally and efficiently consolidated. . . . Furthermore, able workmen were forever breaking away from their employers in Birmingham and setting up for themselves, compounding the fragmentation of work there. . . . To try to describe Birmingham's economy then (and now) is not easy. It was a muddle of oddments. (1969, pp. 87–88)

Jacobs is contrasting two types of competitiveness, which are described more fully in chapter 11. Efficient production comes about when the division of labor in the region creates specializations that can trade successfully with other regions. Flexible production is built on adaptive innovation and improvisation, which come about as the variety of skills in an area become increasingly sophisticated, firms in different industries become more competitive, and the economic structure becomes more diverse. A city may or may not be able to sustain either type of competitiveness. Birmingham's economy remained flexible and adaptive, whereas Manchester's became inefficient. Jacobs would argue that flexible production is easier to sustain than efficient production, although this hypothesis has been neither tested nor qualified carefully.

Jacobs's view of innovation is much like an evolutionary form of social learning, compared to Schumpeter's revolutionary episodes (see chapter 9). People improvise incrementally to overcome economic problems. Economic development is more likely to arise from small-scale innovations and improvisation than from specialization. The development process is cyclical, generating "chain reactions;" development occurs in spurts. Cities develop by replacing imports with local products that are cheaper or better (Jacobs 1985). Jacobs calls such places *active* cities compared to *passive* ones, which essentially are controlled by external forces. Import replacement expands city markets and further stimulates imports from other cities. It also helps diversify the local job base, improve local technology and producer services, expand local capital, and spin off older forms of work to peripheral areas.

Jacobs's description of import replacement is different from import substitution. Import substitution occurs when a commodity previously imported is now produced locally. No modification in the product or production process is necessary to substitute for imports. Import substitution may result in less interregional trade. Import replacement results in new or better products that better satisfy local tastes or that better utilize local technologies. Import-replacing industries can become export industries. Jacobs believes that, for import replacement and mutually beneficial trade to occur, cities at roughly the same level of development must trade with each other. She would discourage diffusion from more-developed to less-developed regions and advocate trade alliances and common markets among neighboring less-developed areas. *Incremental* improvement becomes much more likely when the commodities traded reflect comparable levels of per capita income and appropriate technology that use indigenous inputs efficiently. *Convergent* development is more likely to result from import replacement.

Although her ideas fall short of providing a coherent theory of development, Jacobs offers a comprehensive, dynamic vision of the development process. She would define economic development as the new forms of work that are associated with improvisation. These incremental innovations usually become embodied in new firms that may gradually grow to form new industrial sectors. The idea that improvisation results in new work forms shifts our attention from firms and industries, as basic categories, to the occupations that dominate particular regions. Wilbur Thompson also focuses on occupations to define regional competitiveness (see chapter 11).

Discussion Questions

1. How does price elasticity change over the product cycle?
2. Assume that you could assemble any mix of industries, occupations, or products in a region in the 1990s in order to build a "seedbed" for new products. Which activities would you want located in your region?
3. The midwestern region of the United States has rebounded strongly in the past decade, especially in comparison to the country's coastal regions. Obituaries written for the industrial heartland appear premature. What explains this turnaround?
4. Are the organization and control of production important determinants of regional growth, or are they simply the outcomes of trade theory-type comparative advantage?

5. Which factors would you examine to determine a region's competitive advantage from the perspective of product-cycle theory?
6. How does this perspective differ from trade theory and location theory?
7. Which economic development strategies would make sense if you worked in a region with competitive advantage in new-product creation, standardized-product production, or neither of these?
8. Much emphasis is placed on swift and effective communication in product-cycle theory. In the 1960s, spatial proximity was required. In the age of telecommunications, is proximity still required?
9. How does viewing firm location as a two-stage process make financial inducements more difficult to defend?

Appendix 8.1

LOCATION THEORY AND APPLICATIONS

As noted in chapter 2, location theory and migration theory are beyond the scope of this book. Both consider location decisions at the firm or household level, whereas the theories discussed in Part II consider the regional economy as a whole. Yet, it is important to briefly consider location theory at this juncture, for two reasons. First, Vernon and other contributors to product-cycle theory are arguing with some of the basic tenets of classical location theory; second, location theory has applications economic developers should understand.

Industrial location theory considers how profit-maximizing firms select their location in space. Some of the literature treats production at one point and examines how competing firms select locations to serve demand in a geographic market area. Other literature assumes that demand is centered at one point and considers how land- and resource-intensive production will be organized in space. The latter is the inspiration for central place theory (discussed in chapter 4). Much of the literature assumes equal access to markets, and focuses on how the location of raw materials, intermediate supplies, or labor should influence the location of the production process. The analysis of comparative costs leads to the identification of the best, cost-minimizing location.

The seminal works in location theory are by Alfred Weber, August Lösch, Edgar Hoover, Walter Isard, and William Alonso. An excellent, brief review is found in Alonso (1964, chap. 4). The optimal location of individual firms, although important, does not address the interaction effects that lead to agglomeration; yet, agglomeration economies, which tend to be found in larger urban areas, provide an important focus for product-cycle theory as well as other regional theories. Locational factors such as deep pools of labor, presented in the chapter 4 discussion of agglomeration economies, are important during the new- and maturing-product phases. Once the product has been standardized, the comparative costs of alternative locations and the logic of classical location theory come into play (see Harrington and Warf 1995).

Moriarty (1980, especially chaps. 5-6) and Herzog and Schlottmann (1991, especially chaps. 11-12) provide useful descriptions of how geographically based factors influence the location of manufacturing firms or their branch facilities. The location process has two stages. In the first stage, the large geographic region is identified, from which the markets served by the facility can be accessed. In the second stage, the specific site and community are selected.

If consumer markets are targeted, access to the relevant wholesaling and distribution centers should be identified. If the product serves intermediate markets, the locations of purchasers need to be identified. The product itself, specifically its weight-to-value relationship, determines the relevant mode of transportation, although often several modes are used to ship one product. With market locations and transportation modes determined, geographic areas can be identified that provide reasonable access to these locations in terms of the transportation costs and convenience of transportation service.

Once acceptable market-serving areas are mapped, the analysis turns to the availability and cost of various location factors. Primary and intermediate inputs include land, labor (in appropriate occupational or skill categories), utilities (natural gas, electricity, and telecommunications), supplies (intermediate inputs and maintenance) and public services (police, fire, and so on). Qualitative factors are also considered: the livability of the area (quality of life for employees and especially for management), the business climate (often emphasizing the posture of local government and the degree of union activity), and environmental issues (both the desire for reasonable environmental quality and the threat of costly environmental regulations). Because capital is rarely accessed at the selected location, local capital costs are not very important.

Traditionally, local property taxes and state income taxes have not been influential in the industrial location process, notwithstanding the emphasis given them by state and local developers. However, with the widespread use of financial inducements and incentives in the 1990s, this situation has changed (Wasyenko 1991). With enough corporate subsidies, it may now be possible to skew locational choices away from optimal locations.[16]

The typical manufacturing facility usually can find several locations that offer good access to markets and to roughly equivalent capital outlays and operating costs of production. The process involves starting with many possible locations and screening out more and more communities and sites until only a small number remain. Often, the site-selection decision resolves to more subtle qualitative factors, since the remaining sites have similar capital costs and can provide the necessary production inputs at similar unit cost. In these instances, personal relationships and effective marketing can be decisive.

This brief description of the site-selection process enables one to reconcile two different views of the industrial location process. Most economists, relying on growth theory and trade theory, argue that firms select profit-maximizing locations as a result of logical calculations and rational analysis; they disparage the role played by economic developers in the location process. Economic developers, on the other hand, can document the ways in which they have facilitated the location process and how effective recruitment and marketing have influenced locational outcomes.

In fact, both groups are correct. Firms usually select a place where they can carry out their operations profitably; if they do not, they may not remain in business very long. However, more than one place usually exists where profitable operations are possible. Economists emphasize the screening process that eliminates most communities and sites. Economic developers focus on the hard work and relationship building that usually becomes telling in the eleventh hour when the final location decision is made.

Notes

1. Parts of this section are drawn from Malizia and Reid (1976).

2. Hoover and Vernon did extensive research on the economic development of the New York metropolitan area. See Hoover and Vernon (1959) and Vernon (1963).

3. An overview of location theory is provided in Appendix 8.1.

4. Beginning in the 1930s, Simon Kuznets, Arthur Burns, and others working at the National Bureau of Economic Research provided evidence that industries go through cycles. They identified industry cycle phases—experimentation, rapid growth, stability, or decline. Product life cycles follow a path that resembles industry life cycles. Analysis at the industry level could bring together the product-cycle concept with the industry cycle evidence to form a stronger argument.

Ann Markusen (1985) presents the profit cycle as a better conceptualization than the narrower product-cycle concept. She views regions as highly varied locations with unique histories. Like Allan Pred (see chapter 4), she sees broad generalizations about regional economies on the basis of geographic location as overly simplistic. Markusen changes the assumptions and broadens the scope of the model, arguing that the focus should be on corporate decision-making behavior, which reacts to numerous economic factors. The level and rate of profit determine the trends in output growth. In product-cycle theory, demand is assumed to be adequate, and production decisions are simple reactions to market demand. Output follows the path of growth, stability, and then decline.

Markusen also criticizes the treatment of market structure in product-cycle theory. In the product-cycle model, free entry and price competition require firms to search for least-cost standardized-product locations; in the profit-cycle model, imperfect competition is the rule. Oligopolistic industries change the spatial pattern of production and employment in response to changes in profits. Markusen empirically examines the profit cycle in fifteen industries and finds considerable variation in the pattern and trends of regional location. For a concise summary of the profit cycle as it relates to the product cycle, see Malecki (1997, pp. 64–65).

5. Standardized-product companies from the less-developed country may, in fact, invade and become dominant in the home market of the innovating country. See Barnet and Mueller's discussion of the last phase of their four-phase product-cycle model (1974, pp. 132–133).

6. The benefits of unionization discussed by Thompson (1965), as presented in chapter 11, are not likely to materialize in the less-developed region. Such regions are carefully screened by companies seeking locations for branch facilities and are preferred precisely because the probability of successful unionization there is extremely low.

7. A 1992 study of the economic impact of Ohio State University on the state of Ohio focuses on the university's growth (income and employment) effects while undervaluing its developmental impacts. This study is typical of many economic impact analyses that restrict their focus to output, employment, and earnings growth while ignoring the innovation process that could lead to economic development.

8. Most of the applications to date focus on the state level and provide various forms of assistance to form networks. For example, see Rosenfeld (1995). These applications assume that interfirm networks are needed and attempt to stimulate their formation. Further discussion is found in chapter 10.

9. This section draws from research conducted by David A. Stebbins and reported in his Master's project paper, "Local Economic Development Planning and Practice Case Study: Buffalo, NY," Department of City and Regional Planning, University of North Carolina–Chapel Hill, 1995.

10. The transition from routine to precision production is not without difficulty. First, considerable international competition exists in the production of capital goods for consumer products such as textiles, automobiles, and furniture. Second, as suggested by Moriarty (1983, 1992), restructuring from consumer-goods manufacturing to capital-goods manufacturing is more difficult for smaller, less technologically sophisticated regions.

11. See Vernon (1966, p. 195).

12. Norton and Rees (1979) note the change in the regional location of the industrial seedbed from the mechanical era, which was based on the proximity of machine tools production, to the current era, during which the electronics sector serves as the spatial anchor. The geographic roots of the current seedbed, however, are less deep, due to global communications and the fact that researchers have much

less loyalty to place than the machine tool craftsmen who lived in the midwestern manufacturing belt. The authors use product-cycle theory in arguing that agglomeration economies tie industries to one place while standardization makes them footloose.

13. Moriarty provides more thorough empirical analysis and more consistent conceptualization than either Erickson and Leinbach or Norton and Rees. He examines the spatial diffusion of manufacturing throughout the U.S. urban hierarchy. Although product-cycle theory provides an instructive point of departure, Moriarty provides a more specific locational model. The analysis emphasizes relationships between average wage rates, agglomeration economies, and urban size. Moriarty argues that larger places offer higher wages and more external spatial economies than smaller places. As a result, small-scale, high-wage, capital-intensive manufacturers seek metropolitan locations. Conversely, large-scale, low-wage, labor-intensive manufacturers find smaller places and nonmetropolitan areas more attractive.

Moriarty suggests a reformulation of product-cycle theory. He identifies that theory with the period of the 1950s and 1960s, when market-oriented multilocational firms followed the locational predictions of product-cycle theory. After 1970, the availability, quality, and cost of labor became more important than markets. Large manufacturers sorted out their production to exploit the relative strengths of large cities and smaller places. Nonproduction workers became increasingly prevalent in large areas as processing, fabrication, and assembly operations grew in nonmetropolitan areas.

14. Malecki (1997, pp. 63–71) provides an excellent summary of the relevant literature. He and other researchers draw on aspects of the product-cycle model to understand the dynamics of technology-oriented regional development. Malecki proposes a model that takes into account "the product life-cycle model and its corollaries, the profit cycle, the innovation cycle and the manufacturing process cycle" (1997, p. 63). Although these related cycles enrich our understanding, they are not specified well enough to suggest many practical applications. Product-cycle theory remains more applicable than any of these more complex formulations.

Goldstein and Luger (1990) identify numerous interacting factors that are necessary for the success of research parks. Some high-tech areas have become innovation centers. The premiere successes generally cited are Route 128 near Boston, Silicon Valley in California, and Research Triangle Park in North Carolina. Each of these areas, however, has a complex array of unique characteristics (Saxenian 1994). Policies based on recruitment and promotion of high-tech industries rely on the assumption that the location of these companies will eventually result in growth patterns characteristic of the premier areas. This assumption is tenuous at best. High-tech manufacturing located in research parks tends to introduce process innovations rather than product innovations. While product innovation is the driving force in the product cycle, process innovation becomes important later, after the product matures. Therefore, many research parks simply attract high-technology manufacturing facilities and do not become vibrant innovative centers.

Thus, high-tech firms do not universally serve as growth simulators. Nor do they represent good examples of development that follows product-cycle theory. Definitions of high-tech activities are often not well specified and are applied loosely to different operations of high-tech industries. Malecki (1981, 1990) argues that the connection between regional development and high technology is not well understood.

15. Given these criticisms, how can product-cycle theory be sharpened to provide better understanding of regional economic development? First, the product must be more clearly defined for the set of industries that are exporting or that have the potential to export. The seven-digit commodity code is too detailed for serious regional analysis; the three-digit or four-digit SIC code should be used instead. All these industries that are exporting from the region should be analyzed and, for simplicity, classified as generating either maturing or standardized products. Because product cycles are inherently dynamic, this analysis should be done annually.

If this analysis of product differences fails to proffer clear distinctions, focused examination of the ownership and control structure of major establishments may be more useful. If the region contains numerous branch manufacturing facilities that produce standardized products, the region is likely to be attractive as a relatively low-cost location. The problems and opportunities facing such regions depend on their ability to offer low comparative costs (absolute advantage). If the region contains major headquarter establishments or R&D facilities, it is likely to have a high-wage, high-skill occupational mix and should function much more like a new-product location. Industry size, ownership and control structure, and major function (R&D versus standardized production) provide powerful insights into the region's competitiveness. These factors are discussed at greater length in chapter 11.

16. The issue of financial incentives received considerable attention in the 1990s. The location of the Mercedes plant in Alabama, at extraordinary cost to the taxpayers of that state, brought national attention to the issue. Further discussion of the issue can be found in the Fall 1994 issue of *Economic Development Review*. And, in May 1996, National Public Radio sponsored a national conference on incentives in Washington, D.C., and published the conference proceedings.

9

Entrepreneurship Theories

There is no clear role for an entrepreneur in the neoclassical theories of chapters 6 to 8. With few exceptions, theories previously outlined and discussed have conceptualized the urban and regional economy as operating according to the behavior of economic actors that include investors, workers, producers, and consumers. The motivations of these actors have been characterized in terms of the broad, well-accepted concepts of utility or profit maximization; consumers and workers act to maximize their happiness, or utility, while producers select inputs, devise production plans, and produce goods in order to maximize profits. Through the invisible hand of the market, and with the assistance of a mythical auctioneer who calls out prices of goods and services, the economy achieves an equilibrium where costs are at a minimum and physical and human resources are utilized efficiently. Although definitions of an entrepreneur vary, if the entrepreneur is someone who sees an unexploited opportunity for profit in the form of a new product or even a new way to produce an existing one, then, from the point of view of traditional theory, there is little for them to do. Prices and technologies are known by assumption in the neoclassical framework; there is no uncertainty or imperfect information for the entrepreneur to act on.

Yet, in the modern capitalist society, the entrepreneurial function cannot be ignored. Success stories abound. Entrepreneurs have accomplished far more than has efficient management of resources and personnel. Neoclassical theory notwithstanding, the existence of a particular class of economic actor, the visionary who attempts to marshal resources in unique ways to bring new ideas for products or services to economic fruition, seems undeniable. Moreover, regardless of traditional economic models that ignore entrepreneurship, stimulating entrepreneurial activity is an important, growing focus of economic development efforts throughout the United States. Claims that local developers should focus primarily, if not exclusively, on promoting entrepreneurship are not uncommon (Shapero 1981, Schweke 1985, Yarzebinski 1992).

Entrepreneurship theories are essentially of two basic kinds. The first consists of those theories that attempt to situate the entrepreneur within a broader conception of economic growth and change. These theories usually take neoclassical economic theory as their point of departure. Some important theorists in this vein are Frank Knight, Harvey Leibenstein, Joseph Schumpeter, and Israel Kirzner. The second kind consists of theories that directly address the factors motivating entrepreneurial behavior. Economic theorists are less dominant in this body of work; psychologists, sociologists, historians, and urban and regional planners have been more important in advancing research on motivation. From the point of view of the local developer, theories of entrepreneurial motivation are more relevant, given that they have direct implications for encouraging or promoting entrepreneurial activity. Yet, theories of the entrepreneur's economic role are also important, primarily because they articulate a fundamentally different vision of the growth and development process.

I. Theories

Entrepreneurship theories share a perspective on development that puts human agency at the center of the development process. In other words, it is people at work who make development happen. The key figures are entrepreneurs or, more narrowly, people who carry out venture-creation functions that, in turn, generate development through the innovation. These theories constitute an important departure from neoclassical economic theory, which assigns little role to the entrepreneur (Casson 1987, Baumol 1968, Kirzner 1973). In the neoclassical model, a firm must choose among alternative values for a small number of well-defined variables: price, output level, and input level. The choice involves the consideration of costs and revenues associated with each alternative set of combinations. The firm decides which of these combinations yields the optimal values in terms of profit maximization. As a whole, the economy tends toward an equilibrium where resources are allocated efficiently. In this context, the entrepreneur has no function.[1]

Economists have addressed this problem in two fundamental ways. On the one hand, many regard entrepreneurship as beyond the scope of the self-contained, well-defined problems that economics attempts to solve. It is hardly possible to study entrepreneurship without raising issues of managerial skill, psychology, motivation, and so on. These topics are inaccessible with the tools of the modern, mathematically oriented economist. The result is that the role of entrepreneurship is ignored. On the other hand, a few economists have introduced the entrepreneur by altering the basic assumptions of the neoclassical model to varying degrees. These theories succeed in finding a role for the entrepreneur in broader models of economic growth and change, although they do not address the thornier questions that are most relevant to local development practice: (1) What determines why some undertake entrepreneurial ventures, and others do not? (2) What determines the degree of entrepreneurial activity across particular places? and (3) In what ways might economic development strategies encourage entrepreneurial activity and success?

This chapter summarizes important ideas from these two primary bodies of work. The views of Knight, Leibenstein, and Kirzner are covered first, thus setting the stage for Schumpeter. His ideas, in particular, are addressed at some length in the third section of the chapter, since he provides the most comprehensive theory of entrepreneurship and development. We then examine research on the factors that influence the motivation of entrepreneurs, particularly studies that attempt to identify the characteristics of entrepreneurial places (for example, Shapero 1977 and 1981, Bruno and Tyebjee 1982, Dubini 1988). Discussion of the application of the theories focuses on business development, the role of small firms in the economy, and the link between firm size and innovative activity. The study of small businesses has become closely associated (even synonymous) with entrepreneurship research (see, for example, Sexton and Kasarda 1992). The nurturing of small businesses has become a popular focus of local economic development efforts. The application section provides a critical examination of the logic behind business development strategies, particularly those targeted at small enterprises.

Entrepreneurs and Economic Growth

Knight (1921) recognizes that economic decisions are made in real time under conditions of uncertainty. Of all decisions, planning decisions incorporate the longest time frame and involve the greatest uncertainty. Large corporate organizations can reduce uncertainty by bringing more elements under corporate control and expanding available resources. Thus, the large corporation may, to some extent, be more effective than individual entrepreneurship.[2] Its effectiveness depends on corporate managers who exercise alertness, foresight, and good judgment in making sound planning decisions. The successful managers become recognized as corporate leaders who bear the costs and reap the benefits of uncertainty.

Leibenstein (1968, 1978, 1987) distinguishes "allocative efficiency" from "x-efficiency." Allocative efficiency refers to the efficiency with which "resources and factors of production are combined to satisfy effective demand within an economy." Leibenstein sees x-efficiency as more important. It is tied to "differential and inadequate motivation and information usage." Leibenstein states that "the simple fact is that neither individuals nor firms work as hard nor do they search for information as effectively as they could." He believes that, for any particular firm or industry, the relationship between output and the application of inputs is indeterminate because firms do not achieve x-efficiency; therefore, the production function is never clearly or uniquely specified.

Leibenstein sees the role of the entrepreneur as providing the motivation to increase x-efficiency in a variety of ways. First, entrepreneurs connect different markets to exploit arbitrage possibilities. Second, they remove market deficiencies by "filling gaps" in the market process. Third, they are input completers, in that they coordinate all the inputs required for production. Fourth, and finally, they create or expand firms as productive outlets (Binks and Vale 1990, p. 36). Leibenstein sees the entrepreneur as the person who has the motivation of personal success or monetary reward to keep pushing to reduce the uncertainties in seeking the most efficient production function.

Kirzner (1973, 1979, 1982), a member of the Austrian school of economics, presents the entrepreneur in the context of the general equilibrium system. Leon Walras, who in the late nineteenth century formalized the system in mathematical terms, recognized the need for an information channel to communicate and coordinate a priori plans between consumers and producers. This communication is needed so that the neoclassical assumption of perfect information can be applied in a meaningful way. Walras devised a mythical auctioneer who, as a public good, perfected markets by providing information to various actors about the plans of other actors. This coordination of a priori plans was sufficient to overcome the ignorance reflected in initial price discrepancies. The effective auctioneer would enable households and firms to arrive at equilibrium prices and quantities that cleared all markets.

Rather than being a mythical figure providing information as a public good needed as a precondition for equilibrium, Kirzner's entrepreneur is a real person who exploits arbitrage opportunities in real time and under conditions of uncertainty and, as a result, moves markets toward equilibrium. Thus, the entrepreneur perfects information flows. In the process, this entrepreneur reduces uncertainty and market fragmentation. The primary entrepreneurial trait required to carry out this arbitrage function is alertness to market opportunities. As a result, entrepreneurial actions eliminate price distortions and move the system toward general equilibrium.

Schumpeter (1934, 1947, 1950), who also conceptualizes the entrepreneur in the Austrian tradition, explains the innovation process much more comprehensively (see section III of this chapter). He identifies entrepreneurs who introduce innovations as the phenomenon that best explains capitalist development as a historical process.[3] Innovation creates the "perennial gale of creative destruction" that increases uncertainty and causes development to be an uneven and cyclical process. More important, entrepreneurial innovation increases the social product. Instability is a necessary cost of continual growth; another is the possibility that economic change will gradually give rise to anticapitalist political and social elements.[4]

Theories of Entrepreneurial Motivation

Distinct, yet related to the problem of how the entrepreneur fits into the development process, are the many theories about the institutional, cultural, and educational factors that inspire entrepreneurs to undertake business ventures. Among the best known is McClelland's (1961) model of *n* achievement. McClelland argues that entrepreneurs are motivated by a need for achievement, not necessarily by other more commonly regarded factors, including monetary reward and the acquisition of power. He bases his thesis on a detailed study of entrepreneurship in the industrialized West versus underdeveloped countries. Much subsequent work in the international development field in the 1960s and early 1970s focused on the different factors affecting the supply and demand for entrepreneurship in industrialized versus less-developed countries (Leff 1979).

The most important studies from the local developer's perspective are those that examine environmental influences on entrepreneurship. In a study of Italian entrepreneurs, Dubini (1988) examines the interaction between personality and location. She identifies three types of entrepreneurs, including self-actualizers, people driven by negative circumstances, and followers of family tradition. After examining the spatial distribution of these entrepreneurs and the characteristics of the environments in which they operate, she argues that the barriers to increased entrepreneurial activity vary both by personality and by place. For example, self-actualizing entrepreneurs may face strong cultural barriers in a parochial region that a family tradition–oriented entrepreneur would find very receptive. This variation implies that local developers should identify local impediments that are, in some sense, personality-specific.

A number of studies attempt to identify generic barriers or, alternatively, the characteristics of receptive environments. Bruno and Tyebjee (1982) propose twelve essential factors for a location supportive to entrepreneurial activity: venture capital availability, presence of experienced entrepreneurs, technically skilled labor force, accessibility of suppliers, accessibility of customers or new markets, proximity of universities, favorable government policies, availability of land or facilities, access to transportation, receptive population, availability of supporting services, and attractive living conditions.[5] Most of these factors suggest that assisting the entrepreneur should involve primarily the creation of a supportive environment, as opposed to direct financial assistance.

Shapero (1981) points out that most places equate economic development with more employment, income, or exports and support industrial recruitment and promotion as the best way to achieve this goal. If successful, Shapero argues this strategy is likely to engender greater costs than benefits. Significant public concessions and inducements, new infrastructure and services, and major growth pressures often accompany new facility location. In the long term, the area may become reliant on a few large establishments and share the fate of those narrowly specialized cities that declined because they were unable to adapt to change. According to Shapero, long-term viability requires resilience: the ability to adapt and change with changing economic forces and trends. Resilient places can adapt to opportunities and rebound from adversity more quickly than places that lack people who possess these talents and resources. To build resilience, developers need to create an environment in which entrepreneurship can flourish. Resilience is built on high-quality factor inputs (talented and skilled labor), business and personal services that support innovation, research and development (R&D) activities, and a physical infrastructure that facilitates communication and networking.

Resilient economies need numerous small firms that may succeed or fail but which, overall, offer more stable growth. Shapero argues that small firms, indigenous or recruited, are more innovative than large ones. The most desirable entrepreneurial firms are small high-tech enterprises because they build a region's skills and human resources. Better-educated workers become more fully involved in the community. Thus, local economies become resilient and diverse as a result of people who are creative, willing to take initiative, and able to assume risk. Resilient, diverse areas can adapt to external forces and adopt new patterns of behavior (see Alchain 1950).

The proper "industrial ecology" for resilient, self-renewing economies is built in stages. Expansion of a high value-added company can often be considered a random event. To survive and prosper, however, such firms need (1) proper financing, which requires educating local lending institutions;

(2) a network of supporting business services—technical, legal, financial, marketing; and (3) adequate physical infrastructure. Over time, diverse economic growth increases local incomes and consumer demand. Deeper markets make the region more attractive to additional new companies (Shapero 1977).

Shapero offers numerous attractive insights about innovation and economic development. He puts entrepreneurs at the center of the development process. Entrepreneurs learn from experimentation and from failure. They are motivated by positive and, more often, negative career events. They usually gain valuable work experience with large, established firms. They often strike out on their own after confronting a crisis, such as losing a good job. Collectively, they can sustain economic development. Yet, Shapero's argument has several flaws. He measures entrepreneurship as the rate of new company formation, but this indicator is far too broad, because it treats all new businesses as if they were entrepreneurial firms. As a result, Shapero does not suggest good measures of the capacity communities have for self-renewal.[6] Shapero assumes that all regions have the ability to recruit new or young companies and can build an industrial ecology, or milieu, that will promote new company formation. These assumptions, however, are not correct for many places, especially smaller cities and rural areas. Furthermore, Shapero never clarifies exactly how entrepreneurship leads to self-renewing local economies. He does not define resilience operationally. He misunderstands diversity when he points to the evolutionary process as an example that "nature" provides economic developers. In nature, diversity decreases as closed ecosystems evolve over time.

Entrepreneurship theories do not provide complete theories of the historical or spatial aspects of development; yet, they offer important insights about innovation and economic development. As development evolves over time, entrepreneurs create something new—new combinations, new companies, or, more generically, new forms of work. Entrepreneurial development affects spatial relationships. Resilience and diversity (even though not precisely defined) are important characteristics of sustainable local economies.

II. Applications

The primary application of entrepreneurship theories at the local level in the United States lies in business development strategies. That this should be the case is not as obvious as it seems. Studies of entrepreneurial motivation suggest other types of strategies as well, particularly in the area of workforce education and training.[7] More important, material incentives must be in place to support entrepreneurship. As a reward for assuming business risk, individuals should, without great difficulty, be able to build businesses and accumulate wealth. Economic developers, for their part, should try to sustain local competition and lower barriers to market entry; they should try to make local government regulation of business effective, yet simple. They should support local tax policies that are fair and not oppressive to any particular industry or group. They should be tolerant of new ideas, support democratic processes, and foster ethical standards that promote individual initiative, responsibility, and honesty. With these fundaments in place, economic developers can turn their attention to business development strategies.

Three different business development strategies are often confused: new business development, small business development, and entrepreneurial business development. The difference between approaches is a matter of the specific objectives involved. Focusing on the development of *new* businesses may be attractive as a way to offset, in part, employment contractions or firm closures. The main advantages of implementing strategies to promote *small* businesses, either new or existing, may be greater economic diversity and stability. The developer may focus attention on new and existing *entrepreneurial* businesses as a way to encourage local innovation. The confusion enters when one realizes that, at some point, *all* existing businesses were new, small, and, at least to some degree, entrepreneurial. The distinctions between strategies turn on definitions of innovation and entrepreneurship, as well as the role of firm size.

New business development strategies attempt to promote the local creation of new enterprises of all kinds. A range of programs have been devised to support start-ups, including technical assistance, marketing networks, financing, and infrastructure provision (Popovich and Buss 1990). The most popular approach is probably the creation of incubators that provide new businesses with basic infrastructure and services at low cost for a limited period.[8] Business incubators for new business development are a useful development tool in the same way industrial parks or spec buildings support the industrial recruitment or expansion process. Whether or not they lead to the creation of enterprises that otherwise would not have been established is open to debate. A confounding factor is that many incubators screen applicants to determine their potential for success (Lumpkin and Ireland 1988). Only those most likely to survive are offered space. Of course, these firms are the most likely to succeed in the absence of incubator space.

The more significant problem with strategies for new business development is that, unlike recruitment or expansion, few new enterprises are likely to add many jobs to the local economy. The vast majority of new businesses fail altogether, while most of those that survive fail to grow large. One study found that, of the 245,000 start-ups in the United States in 1985, 75 percent of the jobs created by these firms in the subsequent three years were attributable to 735 firms, a very small share of the total cohort (Harrison 1994).[9] The fact that so few new businesses have significant growth potential suggests that new business development may be a strategy doomed to yield low returns. Yet there are other benefits beyond job creation. New business development may be an important ingredient in neighborhood revitalization efforts, where the scale of the success is in line with the size of the community. Indeed, the proliferation of microenterprise loan funds attests to the demand for start-up assistance in lower-income communities. Nevertheless, the limited growth potential of most new businesses does not bode well for the developer who attempts to use new business creation as a mechanism for increasing jobs significantly.

Small business development and *entrepreneurial business development* are concerned with both new and existing businesses, yet one can safely say that the roles of small businesses and entrepreneurial companies in the U.S. economy are significantly different. Unfortunately, sharp distinctions are rarely drawn. During the 1980s, the entrepreneur became a popular, even heroic, figure. At the same time, David Birch (1979, 1987) presented the idea that new and expanding small firms were primarily responsible for net new job creation. Soon, the culture of entrepreneurship became entangled with the demography of businesses. As a result, small business development, which almost always refers to forms of nonentrepreneurial activity, became confused with entrepreneurial business development. The few young businesses with growth potential, all of which began small, may be considered entrepreneurial firms. Many other businesses fail within a few years. The survivors individually offer few jobs over time.

The terms *small* and *business* are almost redundant. The most reliable statistical information available from the Census Bureau suggests that more than 15 million business entities exist in the United States and almost all of them have only a few employees. About two-thirds have no paid employees at all. These firms typically represent sole proprietorships that provide some personal income but limited or no full-time employment opportunities. Even if attention is focused on firms with paid employees, less than 2 percent have 100 employees or more. In 1987, the average size of companies with employees was 19.2; the same average size existed ten years earlier. The median employment size in 1987 was fewer than 5.

What is noteworthy is the economic importance of the few companies with 100 employees or more. In 1987, these firms controlled 57.5 percent of total employment and 60.4 percent of total sales. These facts do not contradict the evidence that Fortune 500 firms are downsizing and generating few new jobs. The largest firm-size categories were less dominant in 1987 than in 1977.[10]

This static picture does not contradict the statement that "small businesses create most net new jobs." The confusion arises by using the wrong adjective—*small*. As noted above, a tiny fraction of firms are responsible for the lion's share of new job creation. We suspect that these job-creating firms are young, entrepreneurial firms with significant growth potential because they serve large markets, develop new products, or introduce new technologies.[11]

Although more research is needed on high-growth firms, one message is clear: job creation efforts that focus on small businesses per se (companies with fewer than 100 employees) are, like new business development strategies, prone to failure. "Small" businesses eliminate almost as many jobs as they create. Almost all survivors that have employees do not grow (or decline) dramatically. Furthermore, because small businesses dominate the local (nonbasic) sector, their multiplier effects are minimal, whereas their expansion potential is limited by the size of the local market. High turnover and low barriers to entry limit the potential effectiveness of assistance to existing small firms.[12]

Entrepreneurial business development strategies are distinguished by the types of economic activities they target. In this sense, they are much more focused than new or small business development programs. Generally, entrepreneurial firms are defined as young enterprises introducing fundamentally new products or innovations. They are often classified as high-technology firms, although they may belong to any industry. The statistical portraits developed by some of the researchers noted above suggest that successful entrepreneurs typically have considerable business experience, either working for others or in previous business ventures. Many have insider information about how to apply new technologies or meet market needs. They tend to be single-minded, narrowly focused, and driven to carry out an idea they believe has great merit. They may be pursuing an opportunity; but, more often than not, they have been driven to innovate as the result of corporate downsizing or some other adversity.

The typical means by which economic developers might directly support entrepreneurs in their area include financing assistance (accessing venture capital); providing technical assistance (locating suppliers or customers or negotiating regulatory requirements); meeting infrastructure needs (incubator facilities); and facilitating networking through the organization of entrepreneurial councils, entrepreneurship forums, monthly breakfast meetings for the CEOs of young growth companies, and other events to provide opportunities for peer support. Of course, the degree to which the local developer can develop viable programs in these areas varies. Improving the supply of venture capital, for example, is fraught with difficulties. More fundamentally, these programs do not address two important concerns.

First, it is difficult to identify entrepreneurs at a point in their venture when they might benefit from assistance. Entrepreneurs are too busy to surface during the early stages of business development. They usually become visible to economic developers when they have become established and need little help. Second, entrepreneurial companies often achieve success by going public or through acquisition by existing corporations. In many instances, capital raised from going public enables the entrepreneurial company to expand in place or at another location. When acquired, the entrepreneurial firm is often merged with or moved to the location of the acquiring company. Thus, typically, entrepreneurial companies, already few in number, have low success rates, come into view when assistance from economic developers is not highly valued, and often leave an area when they are about to experience significant employment and sales growth. Like new and small business development strategies, entrepreneurial development programs, on average, are likely to have limited impact on local employment. Like industrial recruitment, entrepreneurial development should be considered a high-risk, high-reward strategy.

Innovation, Entrepreneurship, and Firm Size

There is recent evidence that seems to suggest that small businesses (defined as those with fewer than 100 employees) innovate at higher rates than their larger counterparts, at least in some industries. Specifically, the differences appear strongest for higher-technology industries. Calculating the number of innovations per employee by industry and taking the difference between large and small firms, Ács and Audretsch (1987) find that small firms innovate at higher per-employee rates in sectors such as scales and balances, computing equipment, control instruments, synthetic rubber, counting devices, and scientific instruments. Large firms appear to innovate at higher relative rates in traditional industries such as tires, chemicals, industrial and food machinery, and ammunition. The authors use the evidence to claim that there are diminishing returns to R&D (see also Ács 1996).

Are entrepreneurial firms, then, the font of innovation? Entrepreneurial firms are innovative, but the degree and significance of the innovations vary tremendously. On the other hand, the small number of firms with more than 100 employees generate numerous innovations. They have the financial resources to encourage innovation by funding internal R&D and supporting new profit centers. They also are able to find innovations recently commercialized and purchase the innovating company or the rights to use the technology. The issue, again, is not firm size; product, management, competition, industry growth potential, regulatory environment, and so forth are more relevant in determining entrepreneurial success.

These points can be illustrated by using the Research Triangle region as a case study. The Research Triangle Park (RTP) is one of the best-known research parks in the world. In the early 1980s, the *Wall Street Journal* editorialized that RTP, along with Silicon Valley and Route 128 in Boston, was one of the more prominent centers of entrepreneurship in the United States; the editorial was wrong.

From its inception in 1959 until the late 1980s, RTP housed largely institutional research. United States and foreign corporations headquartered elsewhere established branch research facilities in RTP. By the early 1990s, RTP had experienced more than $2.0 billion in development and provided for more than 35,000 employees in a metropolitan area of about 1.0 million population. Over the years, corporate R&D conducted in RTP has served the parent company. Activities range from experiments that have led to new products (for example, drugs to retard AIDS) to routine research that enables the company to customize an existing product in order to provide better customer service. Although IBM put RTP on the map in 1965, eventually employing more than 10,000 people in production as well as R&D, most other major corporations limited RTP operations to research and prototype development.

IBM's R&D activities were innovative to the extent that R&D led to new products and technologies. As a major focus during the 1980s, the company developed, tested, and produced personal computers in the RTP. IBM developed this product line in order to gain a share of a market that was growing much faster than the mainframe market it already dominated. Firms like Apple, Intel, and Compaq were more innovative; IBM was a successful imitator.

In fact, the IBM case seems to illustrate fairly well Schumpeter's argument about structural change. Large oligopolistic firms that have come to dominate a market find their competitive position undermined not by competition from within their own market but by firms producing a fundamentally different, yet substitutive product. In this case, personal computers effectively undermined the market for mainframes.

The most successful innovative company in the Research Triangle area that actually followed the scenario of an entrepreneurial firm is SAS Institute, Inc. Like most start-ups, SAS could not afford to develop a facility in RTP. In the late 1960s, two members of the faculty at North Carolina State University commercialized software products they had developed to implement canned statistical procedures on mainframe computer systems. By the late 1990s, SAS had become one of the largest independent software companies in the United States, with annual domestic and international sales in excess of $300 million.

In the late 1980s, major tenants of RTP began downsizing in response to growing competitive pressures. IBM changed its time-honored policy of job security and began to lay off many employees. Two British pharmaceutical companies merged into Glaxo Wellcome, leading to a merging of R&D facilities and personnel in RTP. Other large corporations had similar, although less dramatic, experiences.

As a result of mergers and downsizing, many highly skilled research and marketing staff were looking for jobs. Some of these workers either left the Research Triangle area or found employment with other local firms. Some became consultants and found enough work to justify staying in the area, at least for the near term. Others founded new companies.

These new companies may be considered entrepreneurial firms. In this case, Shapero's insight that entrepreneurship is encouraged more by adversity than opportunity appears to be correct.[13] Some of these firms are engaged in exciting new-product development in the areas of computer software, genetic engineering, and medical applications. Other founders have created a profitable niche providing services to their former employers. The most prominent among these in the Research Triangle area are contract service organizations that test new drugs and report research findings to major pharmaceutical companies. These activities represent an example of outsourcing, which is a way to reduce labor costs more than a means to spur innovation.[14]

The Research Triangle area in the 1990s has, in fact, finally achieved what was proclaimed a decade earlier: the area has become a hotbed of entrepreneurial activity. The local Council for Entrepreneurial Development is one of the largest organizations of its kind in the country. Experienced businesspeople are coming to the area as never before. Sometimes their motivation is to work in a region where their families can experience a higher quality of life. In other cases, it is to pursue opportunities with growing entrepreneurial firms in the area.[15] New business licenses and incorporations are being issued at rates far above the historic averages.

It is doubtful that local economic developers could have directly assisted these entrepreneurial companies in useful ways. Entrepreneurs appear better able to find support either through spontaneous peer groups or through formal organizations they control, such as the Council for Entrepreneurial Development. As noted, successful entrepreneurial firms are often acquired about the time they become significant employers. Usually they relocate to the headquarters of the parent company, which is seldom in the same area.

Viable Entrepreneurship Strategies

The founding of Columbus on the Scioto River was the result of land speculators bidding successfully for the new state capitol site. The project was not an example of innovation, unless building a new town is viewed as an innovative undertaking. In the 1880s, Columbus developed a specialization as a transportation center. Carriages and wagons were made by several dozen companies of about 100 employees each. Improvisation and problem solving led to a viable sector by the close of the century. Although it is not clear whether manufacturing growth was innovative, it certainly led to a rather diverse manufacturing base that was able to withstand the decline of the carriage and wagon sector in the early twentieth century.

The Ohio State University, founded in 1873, has grown to be a major research university with more than 100 academic fields. Battelle Memorial Institute has been an important source of discoveries since its establishment in 1929. However, the innovations spawned by Columbus-based research have not been captured locally. For example, the Columbus economy is not the sole beneficiary of xerography, which was developed at Battelle. However, applied research is an important part of Columbus's industrial ecology or milieu. Research activities—along with banking, insurance, and a host of business services—provide a rich network of support for local entrepreneurs.

Over the years, this milieu has facilitated innovation in Columbus. Columbus-based entrepreneurs have possessed the ability to anticipate demand, and their companies have introduced successful products and processes in a variety of sectors. Ross Laboratories, for example, can claim among its innovations the heat pump and the freeze-drying process. Les Wexner created The Limited and several other specialty apparel companies that exploit just-in-time inventory systems and other communication systems through stores and mail-order channels. Other examples of home-grown entrepreneurship are found in the areas of fast food (Wendy's), technical information services (Chemical Abstracts), and specialty steel products (Worthington Industries).

Although these and other success stories amply demonstrate that Columbus has been an attractive place for entrepreneurship, it remains unclear how Columbus economic developers can facilitate the entrepreneurial process today. Learning about all budding businesses is impossible; picking the winners from among the known starts is, to say the least, extremely difficult. Financial assistance or guidance in business planning may help people become self-employed, but they rarely serve entrepreneurs who have the experience and skills to build a growth company. As a consequence, many researchers advocate focusing on the local business *environment* rather than on the business. Instead of providing direct assistance to the would-be entrepreneur, developers can attend to the network of business services. Shapero was particularly keen on educating bankers to evaluate new business loan requests more accurately.[16]

Another place-oriented strategy for economic developers who believe that their area can compete by becoming more resilient and diverse is to improve the area's quality of life. Although prone to failure in their early attempts at business creation, entrepreneurs are among the small group of economically active people who can choose where they live. By increasing the attractiveness of Columbus as a place to live, economic developers stand the best chance of attracting the entrepreneur on the verge of initiating a successful business venture. Quality of life is in the eye of the beholder, but most would agree that good schools, safe streets and neighborhoods, affordable housing, convenient access to jobs, shopping, recreation, and reasonable choice in health care, entertainment, and personal services make for an attractive community. To pursue quality in these realms, economic developers would be advised to form alliances with local educators, public managers, and city and transportation planners.

In addition to quality of life and business service networks, economic developers can work to build an attractive industrial ecology for entrepreneurial development. The approach we suggest is to identify sources of localization economies that support existing specializations, then determine whether these sources also have the potential to support new entrepreneurial businesses in other sectors. For example, the presence of the machine tools industry—the sector where machines are designed and made—historically has served as a major magnet that anchored manufacturing in the midwestern "manufacturing belt." Today, a broader range of "developmental services" (discussed in chapter 11) support innovation. A regional economy with the size and diversity of Columbus probably has the range of developmental services needed to form an attractive milieu. Developers should identify these services, consider ways to improve their quality, and advertise their availability to businesspeople.[17] The most attractive strategies based on entrepreneurship theories do not attempt to support innovation directly, due to the inability to forecast markets accurately and the uncertainties inherent in new enterprise development. Furthermore, direct incentives or financing to encourage innovation through new business development are likely to increase competition for the same market opportunities, thereby reducing the chance for any new business to gain monopoly profits. It is preferable to take a long-term approach and gradually build an infrastructure that makes the region more attractive to both entrepreneurial firms and households.

III. Elaboration and Criticism

Although Schumpeter ignored the spatial dimension, his theory provides the most complete picture of entrepreneur-led economic development. The disequilibrium model he developed in 1934 has two circular flows. The upper flow contains routine production and consumption of the economy. Over time, the upper flow reproduces the economy and seeks equilibrium. The lower flow disrupts the upper flow's stationary state through innovation. Thus, only in the absence of economic development and innovation is general equilibrium possible.

Revenues received in the routine circular flow are needed for existing inputs and are not available for risky investments; innovation must be financed from outside the upper circular flow. Bankers (more properly, venture capitalists or merchant bankers) stand ready to provide credit to entrepreneurs from the savings of property owners. By using their power to create new money in the form of credit, they thus provide the funds needed.[18] Due to full employment, credit creation generates inflationary pressures in the economy, which reduce real wages and force saving. That is, because inflation increases profits and reduces real wages, consumption declines relative to total income and therefore real saving increases.

Capitalism, as a process, cannot remain stationary. According to Schumpeter (1950), through innovation, producers anticipate and produce what consumers need, rather than respond to consumer sovereignty. Entrepreneurs are motivated to introduce new combinations—new consumer goods, new methods of production or transportation, new markets, new sources of supply, or new forms of industrial organization created by capitalist enterprise. New combinations are the driving force behind economic development.[19] The creation of new combinations is the function of Schumpeter's entrepreneur; thus, his entrepreneur is neither capitalist (owner), inventor, nor manager. Schumpeter's entrepreneur is the heroic innovator who requires intuition, the "capacity of seeing things in a way which afterwards proves to be true" and of "grasping the essential fact" (1950, p. 85).

The established firm is more likely to use existing methods and procedures to expand market share, thus increasing economic growth. On the other hand, entrepreneurs tend to form new firms that eventually displace older ones. They create new products or markets, find new methods to produce and market commodities, or introduce new methods for organizing production. Through the innovation process, entrepreneurs foster economic development.

With this conceptual model, one should not look at capitalism in static terms nor worry about existing structures of the economy. The more important issue is how capitalism creates structures and destroys existing ones. The capitalist system is "incessantly revolutionizing the economic structure from within, incessantly destroying the old one, incessantly creating a new one" (Schumpeter 1950, p. 84). This process of "creative destruction" or innovation is the essential dynamic of capitalism. Competition from new combinations, then, is the real threat to existing firms, not only in fact but also as an ever-present threat. "It disciplines before it attacks" (Schumpeter 1950,

p. 85). Monopolies and other forms of imperfect competition are not only temporary market conditions, but can be beneficial if based on price-lowering economies of scale.[20]

Schumpeter believes that innovations or new combinations are not evenly distributed over time, but instead appear "discontinuously in groups or swarms" (Schumpeter 1934, p. 223). This process takes the form of a cycle when entrepreneurs, financed by bank credit, introduce innovations. If these innovations are successful, imitators follow, any existing monopoly ends, investment for related innovation increases, and the economy begins an upswing. The railroad and the automobile provide examples, as their introduction caused numerous related innovations. Investment and innovation eventually slow down, however, and mass production, due to standardization of the innovation, floods the market and dampens prices. Rising costs and interest rates reduce profit margins, the economy contracts, and recession ensues. Sequenced around the overall upward trend of the economy, recessions and depressions are a natural response to the grouping of innovations during the previous economic upswing. Business cycles are inherent in the long-term process of economic growth.

In his view of a developing economy, Schumpeter believes that above-normal profits are the reward for innovation and the assumption of risk. It is the price society must pay for revolutionary gains in economic development. In addition, Schumpeter contends that capitalism actually reduces inequality in several ways: (1) by increasing equality of opportunity relative to earlier, more class-bound societies, (2) by creating mass-produced products that benefit the working masses more than they do any other sector of the economy, (3) by philanthropy and social legislation underwritten by the process of capitalist economic growth, and (4) because "absolute poverty" falls as capitalist development proceeds, although the variation in incomes and wealth may increase.

In summary, the entrepreneur investing under risk and uncertainty is not the rational calculating manager making routine decisions. The entrepreneur is neither manager nor inventor nor capitalist. The capitalist may risk funds, but the entrepreneur controls the use of those funds. The entrepreneur is an innovator who introduces new combinations—a new product, new method, new market, new resource, or new organizational form.

The introduction of new combinations, which are spontaneous and discontinuous, lead to cyclical development and the "creative destruction" of existing activities for new and better ones. In this vein, Schumpeter is not concerned with the monopolization of markets or the rise of colluding oligopolies. Large producers often are more efficient than smaller producers, and oligopolies will ultimately be smashed by innovation that makes these industries obsolete. Of course, this position is supportable as long as entrepreneurship continues. Dominant companies have been able to purchase inventions and suppress innovations that threaten their profits and market share. Large corporations may also be better able to sponsor R&D and introduce new technologies.[21]

To his credit, Schumpeter assigns an important role to development finance. Again, his ideas must be reconsidered in the contemporary world of financial sources and instruments. Corporations can finance innovation from retained earnings (business savings) or from the sale of securities (stocks and bonds). Independent businesspeople must rely primarily on commercial lenders. But these financial intermediaries are not often sources of development finance (equity capital). Independent businesses need access to long-term equity capital, what is called "patient money," that is not easily found. Most new businesses continue to be financed by informal sources and personal loans (credit cards or home-equity loans).

According to Schumpeter, not only does capitalism produce commodities, it also produces a culture. As rationalization becomes dominant, capitalist development begins to undermine the institutions of property and contract. Socialism increasingly becomes possible. Schumpeter was not completely serious with this analysis; instead, he was preaching doom to thwart the threat of socialism he perceived after World War II.

In regions that are largely technology-borrowing and import-substituting, some economists have claimed that Schumpeter's priorities of innovator over "mere manager" should be reversed. One of the problems lies in the inefficiency of the routine managerial functions that prevent entrepreneurs from continuously expanding their firms and from moving into more complex economic activities.[22] In addition, the entrepreneur does not have to stimulate demand; imports have already mapped out large markets. In this environment, perceiving truly new economic possibilities and pioneering technical and organizational innovations are not very relevant.[23] The work of Kirzner (1979), Baumol (1983), Leibenstein (1987), and Kilby (1988), among others, illustrates the need for efficiency-seeking behavior. This need includes gaining the ability to manage an increasingly complex business organization.

Theories of entrepreneurship never quite articulate the stimuli for innovation. Often, innovation comes about because of time and space constraints. The introduction and widespread use of facsimile machines was a response to time constraints and the need for productivity improvements in the information-processing service industries. Higher-income households increasingly face severe time constraints, thus are willing to spend more money on time-saving goods and services in the areas of meal preparation, housekeeping, child/elder care, fitness and recreational activities, and entertainment. The miniaturization of products through application of microelectronics, even loft apartments in dense urban areas, are a partial response to space constraints. Information-storage technologies and just-in-time inventory systems share the inspiration of saving both time and space.

These points suggest the need for further research in entrepreneurship, especially research that can help increase the odds of entrepreneurial success in particular locations. Yet, as they stand, theories of entrepreneurship provide a substantial counterweight to economic growth theories. The latter analyze the economic outcomes of existing structures, the former how the economic structures themselves change. This distinction is an essential aspect of the difference between growth and development.

Discussion Questions

1. What are the benefits of focusing on the development of *new* businesses?
2. What are the advantages of implementing strategies to promote *small* businesses, both new and existing?
3. For what reasons might the developer focus attention on some definition of new and existing *entrepreneurial* businesses?
4. What social or cultural factors determine the inclination to innovate?
5. Does regulation—the general extension of state control over the economy—reduce or motivate innovation?
6. Do entrepreneurship theories offer insights as to how the goals of sustainable environment and sustainable economy might be balanced? Is it possible to undertake ecologically sustainable development while also ensuring economically sustainable development?
7. Can the entrepreneurial element of economic advance be promoted directly, or is it better to improve livability and the business climate for entrepreneurs?
8. Give examples of ways an economic developer can "particularize" alienation theory to better understand the local economy.
9. Does the concept of exploitation provide a compelling explanation of undevelopment in less-developed countries or in declining areas of the United States?
10. Are regional wage differentials better explained by productivity differences or by historical/cultural factors?

Appendix 9.1

MARXIST ANALYSIS AND THEORY

The best way to summarize Marx's theory of political economy is to contrast it to neoclassical theory. The latter is based on an approach to science called "positivism," while the former is based on materialism, modified by the adjective *historical* or *dialectical*. Positivistic social science aims at explaining, predicting, and ultimately controlling physical and human behavior. To this end, behavior is specified with quantitative variables and causal relationships. Under positivism, scientific progress is achieved when research results can be generalized to build more universal theory.

In contrast, Marxist theory is *critical* in the sense that it attempts not only to understand the world, but to understand it in order to change it. The purpose of theory is to get beyond appearances to the essence of things. Social progress is made when the general comprehension of reality leads to specific action that achieves desired change. Rather than generalization, particularization describes the direction of inquiry. Through specific action, theory is used as it is tested. The success of these actions reflects the veracity of the theory.

As a positivistic science, neoclassical economics is concerned with an abstract theoretical model of an equilibrium system from which specific hypotheses about the economy are drawn. Marxist political economy consists of historical, holistic, and synoptic thinking intended to understand how conflict leads to change and flux. In other words, to understand development in one place, we must study the global economic system; to understand the economy, we must examine the political and social spheres along with the economic sphere. The holistic perspective encourages the search for connections between institutions and production. Marxist theory encourages us to understand the institutions of private property, legal contracts, competitive markets, and regulation of the supply of money, labor, and land in order to grasp their connections to production. The theory's historical perspective forces us to address the question: how did things get the way they are? Thus, to understand the economic history of a place, we have to study real people and events over the relevant period of time.

The unit of analysis in Marxist analysis is particularly difficult to grasp. Traditional economic theory analyzes such units as households, firms, or industries, considering them to be separable and additive. In the Marxist tradition, the individual is viewed as part of social relations; thus the social relation is the unit of analysis. Parts of the social relation are bound together and are mutually causal. In microeconomic theory, in order to behave as utility or profit maximizers, individuals are separated into consumers or producers. In Marxist theory, whole people in reciprocal relations influence the development process. Finally, a static human nature underlies the neoclassical model. In Marxist theory, human nature is unfolding or continually changing; therefore, human nature is dynamic.

Marxist theory is built on dialectical laws. However, the explanation of this dialectic is complicated, because it serves as an image of reality, as a method for analyzing reality, and as a form of exposition.[24]

Alienation

Marx's major work, *Capital* (1967 [1867]), is a study of alienation under the capitalist mode of production. Instead of serving as the point of departure for a mediated theory of human action (praxis), Marx's theory has been interpreted as a closed system of thought driven by technological determinism. Without action as a way to test hypotheses, such thinking remains abstract, closed, static, and subject to misuse and abuse.

Alienation is the general theory from which the theory of capitalist political economy is drawn. Alienation is an unnatural separation or division; it exists in all areas of life because producers are separated from the means of production, which results in, and is sustained by, class society. Unequal power relations make things (objects or concepts) more important than people.

At the root of this analysis is the study of class relations. Class analysis may seem passé in the high-tech, postindustrial era. It is conspicuous by its absence from most recent neo-Marxist work. Multiculturalism has made us aware of the many personal characteristics that make us different: gender, race, age, sexual preference, color, ethnicity, religion, physical size, and capabilities. Neglected in this thinking are ways to build on our similarities. Instead, we recognize our differences and compete in the political arena for limited resources. From a Marxist perspective, such interest-group competition diverts talent and energy away from real economic issues that could lead to meaningful change and toward pseudo-political issues, which result in just enough change to keep all essential relations intact. As long as they identify with or distinguish themselves from others on the basis of gender, race, age, and so forth, no one will think about how they can, as a member of a class, achieve economic development for themselves and others.

In summary, Marxist theory treats the relations of production as basic units of analysis. The definition of economic development is broadly conceived as the reduction (and eventual elimination) of alienation. People organized into social classes take action to influence the economy, which is part of the larger social reality. At its core, Marxist thinking is based on the eighteenth-century assumption that human beings can improve as a species, and therefore its view of secular progress and human nature is optimistic. If, on the other hand, we are hopelessly uncivilized brutes, always requiring authority to keep the peace, then the Marxist believes that we will perpetually suffer alienation from our products, our tools, our natural environment, and one another.[25]

Regional Theory

The Marxist perspective highlights the interdependence of more-developed and less-developed regions (and of metropolitan centers and their peripheries). Economic exchange between such regions is inherently unequal because domination is an essential feature of capitalism; the capitalist class in the more-developed region exploits all classes in the less-developed region. In the production process, workers who are fortunate enough to be employed are exploited to the extent that they generate surplus for the owners of capital. At the aggregate level, less-developed regions are exploited to the extent that the net flow of this surplus moves in the direction of more-developed regions. People in these less-developed regions are better off than those in regions that cannot participate in the larger economy, because the latter have nothing to produce and sell competitively. The most marginal regions play a role that is similar to the "reserve army" of unemployed workers.

Marxist theory provides a useful counterpoint to neoclassical theory by arguing that unequal exchange is an inherent feature of capitalist development. Unlike the logic of trade theory, greater productivity cannot be achieved without greater inequality. The economic system, then, does not consist merely of more-developed and less-developed regions. This dichotomy conveniently ignores economic history. Rather, the former have become more developed at the expense of the latter; development in one place leads to underdevelopment in another.

To explain the evolution of capitalism, Marx presented the essential idea that the class that owns capital exploits labor.[26] Subsequently, through technological progress, capitalism is able to create more wealth and productive capacity than any previous mode of production in human history; the gap between production for use and production for the market widens. Greater efficiency and labor-saving technologies, however, generate overproduction and unemployment, leading to increasingly severe economic crises and eventually enabling the working class to expropriate the capitalist class.[27]

Marx saw capitalism caught in these contradictions, which would spell its doom in spite of efforts to save the system. One way to postpone these crises is through imperialism—the expansion of the capitalist system through foreign trade and direct investment. Trade to other regions is attractive when products can be sold at profitable levels in these markets. Investment in other regions is attractive when wages are lower and other inputs are cheaper. Foreign trade may postpone realization crises; foreign investment may postpone liquidation crises.

Interregional Wage Differentials

The historical perspective can also be usefully applied to labor market relations in different regions of the world. This view is contrary to neoclassical theory, which assumes that all labor is paid its marginal product. Wages are *historically* determined and exogenous to labor markets where supply and demand interact. At any given time, the average laborer is compensated to the extent necessary to achieve generally expected and acceptable levels of living. These expectations are formed by workers' previous experience, comparisons to peer workers, and what is known about the experience of previous generations of workers. Far from being separate activities, workers' wage expectations are based on a direct connection between consumption and production. Workers seek employment in order to earn acceptable levels of living, and the levels of living experienced reinforce the need to work. Development occurs as levels of living gradually improve: expectations rise, compensation demands increase, demands are met with higher wages, and living levels improve.

The different cultural experiences among workers in different parts of the world account for the existence of vastly different compensation for essentially the same work. Workers tend to get what they find acceptable, regardless of their average or marginal productivity. Within the United States, more modest differences in expectation exist. Still, workers in small towns or rural areas tend to expect less than urban workers. Immigrants usually are willing to take jobs shunned by domestic workers. Neoclassical theorists have tried to explain away these differences by controlling differences in the cost of living and in educational levels between people or places. What is ignored (and is much more important) are the differences in the composition of consumption. Workers with simple tastes and lifestyles cost employers less than sophisticated ones.[28]

Notes

1. Citing Baumol (1968), Casson notes:

   The "disappearance" of the entrepreneur is associated with the rise of the neoclassical school of economics. The entrepreneur fills the gap labeled "fixed factor" in the neoclassical theory of the firm. Entrepreneurial ability is analogous to a fixed-factor endowment because it sets a limit to the efficient size of the firm. The static and passive role of the entrepreneur in neoclassical theory reflects the theory's emphasis on perfect information—which trivializes management and decision-making—and on perfect markets—which do all the coordination that is necessary and leave nothing for the entrepreneur. (1987, p. 151)

2. Because the global economy is dominated by the large corporation and not by independent firms, it is relevant to examine corporate innovation. One hypothesis is that the business groups that are organized within many countries can provide the entrepreneurial function (see Strachan 1976). Another hypothesis is that through "intrapreneurship" individuals can innovate within the corporate umbrella (Pinchot 1985). In either case, identifying innovators and understanding the innovation process in the U.S. context is critically important.

3. Schumpeter's theories are based on assumptions of an existing capitalist institutional environment with private property and initiative, a monetary and banking system, a spirit of industrial bourgeoisie, and a scheme of motivation characterized in the nineteenth century by advanced economies (Schumpeter 1934, p. 145).

4. Schumpeter acknowledges that cyclical unemployment and economic inequality will occur, but he believes that the long-term trend will be positive. Contrary to Marx's prediction, unemployment as a percentage of the labor force has shown no upward trend; the capitalist system will continue to be productive enough to provide relief of cyclical unemployment (Schumpeter 1950). See Elliot's (1980) comparison of Marx and Schumpeter; also Elliot (1985) and Appendix 9.1, which is on Marxist theory.

5. The factors listed are similar to the traditional location factors cited in the previous chapter and cited in Appendix 8.1. The first six listed here appear to be the most important from the perspective of the entrepreneur.

6. Possibly, the survival rate of new companies or their diversity or the capacity of young firms to export would be more useful measures.

7. Microenterprise loan programs, for example, use training and peer support groups in an attempt to create local entrepreneurs (although evidence suggests that most people who participate in these programs tend to establish retail or personal-service businesses with minimal growth potential). In general, younger persons should be able to pursue careers locally rather than migrate elsewhere to find economic opportunity.

8. Incubators typically provide low-cost space, shared office assistance, and business development assistance (Allen and McCluskey 1990).

9. Harrison's (1994) numbers are from an internal study by Dun and Bradstreet.

10. These figures are from the U.S. Census Bureau's *Enterprise Statistics.* The

Office of Advocacy in the Small Business Administration has compiled enterprise information from business census sources for 1988 through 1994 that generally corroborate the 1987 results.

11. In other words, the employment size of high-growth firms is an unstable category. For example, a high-growth firm may have fewer than 20 employees for several years until it enters its high-growth stage. Within a year or two, it may have more than 100 employees. Thereafter, it may enter a maturing stage during which employment growth occurs gradually. As a "small business," it experienced highly variable rates of employment growth. Focusing on a firm's transitory employment size is both confusing and irrelevant.

12. In addition, wages, hours, working conditions, fringe benefits, and advancement potential typically are better in well-established corporate enterprises. Further discussion of small business development strategies can be found in the winter 1992 edition of *Economic Development Review* (vol. 10).

13. In interviews and presentations, David Birch often identifies creatures in the animal kingdom to distinguish among the firms he tracks. Most firms are mice; they are small at birth, remain small, and have short lives. In comparison to the multitude of mice, there are a few elephants; elephants grow to be large and live for a long time. The few rapidly growing firms—the gazelles—survive through alertness and speed. Gazelles appear to illustrate Shapero's research on entrepreneurs. They are motivated primarily by insecurity and live in constant fear of being eaten by large felines (what has been called "entrepreneurial terror"). Economic developers also want to catch some gazelles; unfortunately, they usually do not recognize them until they go speeding by, heading for greener pastures.

14. The high cost of permanent employees has motivated many companies to limit such hires and fill many positions, especially entry-level, high-turnover ones, with temporary employees. Companies can hire these employees at attractive salaries and still save significant sums by not paying social security taxes, health insurance, retirement packages, or other fringe benefits.

15. For example, one entrepreneur who founded a software company near RTP in 1995 did so because he believed he would be better able to attract venture capital with an RTP address.

16. This program, however, is not likely to succeed because banks are not a reasonable source of patient money or risk capital. The best way to finance innovation is the subject of long-standing debate. Options include relaxing regulations on financial institutions, creating new secondary markets for small-company securities, changing the taxation of dividend income and capital gains, and providing tax credits for qualified investments in young companies.

17. Economic developers in Columbus seem to recognize, at least partially, the importance of developmental services. *Business First* (Top Twenty-Five Lists, October 26, 1992), a business newspaper for the Columbus area, lists the top twenty-five companies in many areas that include important developmental services—for example, legal, accounting, advertising/marketing, architecture, banks, employment agencies, engineering, and commercial real estate developers.

18. The neoclassical growth model ignores the money sphere by assuming that

real saving will finance direct investment conducted with perfect foresight. Schumpeter sees the money sphere as central. As Austrian finance minister after World War I, he was well aware of the role of finance in the economy.

19. Schumpeter defines development as "a distinct phenomenon, entirely foreign to what may be observed in the circular flow or in the tendency towards equilibrium. It is spontaneous and discontinuous change in the channels of the flow, disturbance of equilibrium, which forever alters and displaces the equilibrium state previously existing" (1934, p. 64).

20. The auto industry, for example, may have offered cheaper and higher-quality cars in the 1920s as an oligopoly than it had offered previously as a more competitive industry. Competition and the advantages it provides are not only found in the market for existing products but in the arena where the threat of new products and industries provides competitive discipline to existing firms.

21. Multinational corporations have an advantage in the area of technological innovation because they have greater experience in the management of research and development, a lower cost of capital, and a lower discount rate for evaluating the present value of R&D (Leff 1979).

22. Kilby (1988) believes that there is an enormous backlog of new technological products and unapplied production techniques, both in the developed industrial economies and in the less-developed economies.

23. To adapt techniques and organization, maximize factor productivity and minimize costs, secure working capital finance, improve substitutes for nonavailable skills and materials—all are tasks on the production side that, more often than not, represent the critical entrepreneurial function in the modernizing economy of the late twentieth century (Kilby 1988).

24. Rather than try to describe these laws, it is more useful to illustrate one of them—the unity of opposites—as a way to explain Marxist theory more fully. Distinctions commonly made in traditional theory are unified by using the social relation as the unit of analysis. In traditional theory, facts are supposed to reflect what is given and values are said to represent what is desired. Yet, these apparent opposites actually are not fundamentally different. Rather, as noted in chapter 2, facts are made as the result of human action. Historical facts (data) are given; future facts (values) are produced as human intentions are realized.

The two-sided, reciprocal approach not only requires explanation of why things are different, but why they are part of the same process. The dualist distinctions between employment and unemployment, affluence and poverty, or development and underdevelopment tell only half the story—and the less important half at that. The more pressing question is: how do these aspects relate as part of the same process? Analysis of essential relations leads to the conclusion that some people become employed or affluent, while others become unemployed or remain impoverished, or that some places become more developed as others become increasingly underdeveloped.

25. For further discussion, see Ollman (1971).

26. To Marx, exploitation is not an idea to be debated on moral grounds; it is a fact that is a necessary condition for production and progress under capitalism. Under feudalism, workers were tied to the soil, but they also enjoyed some degree

of security. Under capitalism, workers are free to seek employment, but they have no direct access to the means of subsistence. Although the capitalist state appears to represent the community, it functions as an instrument of class rule to maintain existing social relations. The state defends legal institutions with its monopoly on the means of violence. Civil law enforces contracts in order to resolve conflicts peacefully; criminal law enforces the rights of individuals, especially property rights. Although governance may be carried out in many forms, unequal relations based on wealth define agendas and influence political outcomes.

27. Realization crises and liquidation crises are described in Bronfenbenner (1979, chap. 5). In a closed system, the process of transforming money into commodities and commodities into money breaks down in the following ways. The amount of capital involved in production continues to rise as labor-saving technology is introduced; consequently, the rate of profit falls. Eventually, a minimum rate of profit is reached that is too low to induce further investment. Capitalists hoard money, which results in underinvestment and a "liquidation" crisis where money fails to be transformed into commodities.

Alternatively, capitalists can sustain an acceptable rate of profit by garnering more surplus from production. The cost, however, is generation of greater unemployment. More unemployment reduces aggregate demand and leads to a "realization" crisis in which commodities fail to be transformed into money. Marx expected liquidation and realization crises to alternate as capitalists try to postpone the system's collapse. Capitalism suffers these crises, which become more and more serious, until revolution occurs.

28. Historical differences in regional culture have directly affected wage levels. Regional cuisines, recreational habits, forms of entertainment, and shared values have tended to reinforce wage levels. For example, Southerners who ate food they raised or hunted, and who spent much of their free time at church, could work for less than others who had more expensive lifestyles.

Entertainment is one of the most successful American exports. It appears to be influencing tastes all over the world, especially among younger people. Over time, more sophisticated tastes will translate into demands for higher wages, which could change competitive advantage and influence international investment decisions dramatically.

10

Theories of Flexible Production

This chapter outlines the major perspectives, theories, and applications of ideas about flexible production. In general, the flexible-production literature focuses the developer's attention on the organizational features of the regional economy—from the organization of production within resident firms to the means by which firms manage their relationships with other enterprises and institutions. It highlights as well the significance of the "embeddedness" of economic relationships in broader social and political contexts. Put differently, the related flexible-production theories demonstrate how social, cultural, and political conditions and relationships can have an important influence on the performance of local firms. In terms of application, such theories suggest that developers should take a coordinating role in setting up formal mechanisms for information and technology transfer between firms (modernization programs and networks), identify and address significant barriers to the adoption of new technology among local businesses (particularly small businesses), and target recruiting and existing industries programs on building local industrial complexes (clusters).

The overview in the first section provides a brief summary of the primary features of the theory and major implications for policy. The next section outlines several applications of flexible specialization ideas, including policies to encourage firms' investment in advanced technology, the development of industry clusters, and the creation of interfirm networks. Section III, "Elaboration," expands on production ideas by first exploring the concept of flexibility itself according to three scales identified by Gertler (1988):

(1) flexibility in machinery, (2) flexibility in the organization of work within the firm, and (3) flexibility at the level of the economy. We then summarize two views of the implications of flexible production for regional economies: the transactions cost approach of Scott and the analysis of new industrial districts. The section also considers areas of recent focus in the flexible-production literature, including questions of labor relations, buyer–supplier trends, and subcontracting practices. Some of these areas of inquiry evolved from criticisms of early (optimistic) assessments of the implications of flexible production for regional growth and change; as such, they have helped foster a more sophisticated understanding of flexible-production ideas.

I. Overview

In 1984, Michael Piore and Charles Sabel argued that the post–oil-crisis industrial economies are crossing a second "industrial divide," a period of profound uncertainty where the direction of technological development is at stake. The first industrial divide, which occurred in the nineteenth century, marked the emergence and subsequent dominance of mass production methods over more craft-based modes of economic organization. Henry Ford's pioneering and widely imitated method of production, which drew its strength from massive economies of scale, standardization, and internal division of labor, as well as large and stable sources of demand, is the oft-cited example of the type of technology that eventually drove smaller, craft-based producers out of business; indeed, the literature characterizes traditional modes of production of this kind as "Fordist." In the current period, however, presumably it is mass production itself that is in crisis, its rigidity and scale insufficient to respond effectively to fundamental changes in the nature of demand. Of course, Piore and Sabel were writing during a time of relative economic stagnation (the mid-1970s to early 1980s), when few major industrialized countries outside of Asia were performing particularly well.

Piore and Sabel described one alternative to the crisis based in Fordist methods as continued reliance on techniques of mass production, but with an extension of Keynesian-type macroregulation to an international level, where the focus is on increasing the purchasing power of consumers in underdeveloped economies (in effect, building markets for mass-produced goods). The other alternative they described as the spread of "flexible specialization," a partial return to a less rigid and more craft-based technological model characterized by a type of cooperative competition among smaller firms, the use of flexible manufacturing equipment and techniques, and a greater reliance on social relationships (for example, trust) as a means of organizing transactions (Best 1990). In illustrating this model, Piore and Sabel cited successful regional economies in the United States and Europe consisting of small, networked enterprises. These firms had found ways to coordinate production and innovation in order to serve rapidly changing niche markets. Also, unlike the branch plants of large multinational companies, they were closely tied to their regions.[1]

The regions where these firms are prevalent have subsequently been viewed as models of development to be replicated in other places. In the late 1980s and early 1990s, a significant share of the research activity in economic development focused on identifying and confirming Piore and Sabel's hypothesized shifts in the basic organization of firms and industries, as well as their implications for local economies. The flexible specialization perspective fueled related literatures on the adoption of flexible manufacturing techniques, the link between industrial organization and agglomeration, and external economies in "new industrial districts" (building on Alfred Marshall's study of industrial districts in the late nineteenth century). Practical applications of the ideas are found in various modernization programs, the development of industry clusters, and the organization of formal, interfirm networks. The view that flexible production is a certain route to competitiveness has become so prevalent in urban and regional economic analysis that some have described it as the "new orthodoxy" (Amin and Robins 1990).

In the early to mid-1990s, research on flexible specialization shifted to an examination of some potential negative impacts of flexible-production regimes as observed in some industrial district contexts. Harrison's (1994) critique of labor practices in many prototypical industrial districts and subsequent researchers' examination of asymmetries of power between buyers and suppliers (often between large buyers and their smaller and dependent subcontractors) helped spur more balanced inquiry of ideas about flexible production.[2] (The debate over the importance of small firms for job creation and economic development is discussed in chapter 9.) Cooperation between contracting firms does not necessarily imply an even playing field between partners, suggesting that to remain competitive, some firms may require development assistance of various kinds.[3] In general, flexible-production ideas challenge the development practitioner to develop a sophisticated understanding, not only of markets and industries relevant to the practitioner's region, but also to internal corporate strategy, interfirm business relationships, and the specifics of production technology.

II. Applications

There are several local development applications and strategies that are informed by theories of flexible production. At the first level of the flexibility hierarchy are initiatives designed to increase adoption of advanced production machinery and techniques (including quality and workforce management) that permit greater flexibility in improving or modifying existing designs, introducing new ones, and maintaining quality. Modernization policy may be concerned less with industrial organizational issues than with the basic factors that prevent firms from investing in equipment upgrades and adopting state-of-the-art management practices. These factors include unavailability of investment capital, lack of expertise, low levels of workforce skill, and difficulties in getting new systems up and running properly. Nevertheless, advanced equipment and practices are crucial to achieving greater flexibility. Therefore, modernization policy has clear implications for achieving flexibly specialized regional economies, if that is the goal.

One of the most important (and, recently, most popular) economic development applications that draws on flexible-production ideas is the focus on industry clusters, a method of understanding and promoting key regional strengths popularized by Michael Porter's (1990) *Competitive Advantage of Nations* and less directly by Piore and Sabel (1984) and the new literature on industrial districts. Closely related to industry cluster strategies is the establishment of formal interfirm networks. In fact, clusters essentially are a network of producers, linked either through formal trading relationships or through shared factors or knowledge. We consider each of these areas of application.

Modernization and Technology Adoption

With the rapid rate of innovation in all spheres of the economy, both state and federal policymakers increasingly have focused their attention on the barriers firms face in upgrading their facilities with state-of-the-art production technology.[4] Initially, attention focused on the rates of adoption of production equipment, but more recently workforce management, quality management, and even buyer–supplier contracting practices have received attention. Much of this focus is on small- to medium-size manufacturing enterprises (SMEs), on the basis that these firms face greater difficulties in obtaining access to, purchasing, and implementing advanced technologies. The federal government, through the National Institute of Standards and Technology (NIST), has partially funded a national manufacturing extension service (Manufacturing Extension Partnership—MEP) that is implemented by state and local government agencies and local nonprofit organizations. Currently, the network of centers extends to all fifty states and Puerto Rico and includes some 300 field offices. Essentially, MEP centers provide consulting services in the areas of quality assurance, production process improvements, information technologies, marketing, design, regulatory issues, engineering, and workforce development.

Industry Clusters

Typical economic development practice focuses on the needs of individual firms, in terms of assistance for locating sites, providing worker training, and constructing and maintaining infrastructure. In Appendix 4.1, we discuss how a growing number of cities and states have begun implementing industrial cluster strategies, which are essentially economic development initiatives targeted to groups of linked firms.[5] In that discussion, we focused on how clusters represent an application of techniques designed to understand industry linkages (input–output analysis, for example). Flexible-production ideas are a major source of the theoretical (as opposed to technical) underpinnings of cluster strategies.

Rosenfeld calls the policy interest in industry clusters "a radical departure from traditional economic development strategy, which, whether aimed at business development or business retention, is always applied firm by firm" (1995, p. 7). Cluster strategies recognize that the fortunes of individual businesses are, in many ways, defined collectively, given that they depend on common factors (inputs and labor), improvements in technology, and the general growth of the economy as a whole. Clusters are also a departure from the standard neoclassical view of market economies, which emphasizes fierce competition between atomistic producers. Traditionally, cooperation has been viewed in terms of price-fixing and collusion and therefore was considered detrimental to the overall performance of the economy. Clusters emphasize cooperation primarily in terms of innovation: a recognition that knowledge spillovers help drive technological advance and, ultimately, economic growth.

In theory, clusters are a geographically concentrated group of firms essentially interdependent along one or both of the following dimensions: (1) presence in the same product (or input–output) chain; (2) important similarities in technology or workforce requirements.[6] Clusters might also be characterized by the presence of related organizations (educational institutions, business associations, formal networks). If they are not, this is often the first point of attack for the local developer, who might try to set up dedicated training programs in community colleges, establish a business advisory group to study ways to identify and solve problems jointly, or look for possible synergies between industry and research and development (R&D) activity occurring in nearby universities. In practice, because of a paucity of good data on local linkages, as well as lack of agreement regarding how best to identify clusters for the purposes of policy, clusters often are poorly defined as regional industrial specializations, and little attempt is made to understand the social, cultural, and political factors and context that influence cluster success.

Feser (1998b) identifies two types of cluster policy applications: (1) those that attempt to build particular clusters as an explicit goal of policy (cluster-specific strategies), and (2) those that use cluster concepts as a way to improve implementation of standard or traditional development programs and initiatives (cluster-informed strategies).[7] An example of the former is an attempt by a community to nurture a specific cluster identified by public officials and business leaders as an existing or emerging specialization. Such policies may include both demand-side and supply-side elements, as public-sector attempts to stimulate demand for the various related outputs of the cluster (perhaps achieved through changes in regulation, focused government purchasing practices, or targeted consumer information programs) are coupled with programs designed to improve the competitiveness of cluster firms (such as technology adoption, business assistance, and networks).

An example of cluster-informed strategies is the use of industry cluster analysis (the identification of different clusters and their member firms in given regions) to better focus modernization strategies. For example, local developers might seek to improve the adoption of advanced, flexible-production technologies among regional manufacturers by focusing assistance and information provision on large, regional end-market producers. To the degree that technologies tend to diffuse backward through production chains, such producers may influence upstream suppliers and subcontractors to adopt compatible machinery and techniques. Information sharing within the cluster (for example, through informal networking) may also help diffuse technologies even between firms not formally linked in the production chain. This is an example of how cluster policy, as a way of implementing flexible-production ideas, can serve to leverage scarce development resources.

Interfirm Networks

As should by now be evident, closely related to the concept of industry clusters are interfirm networks. One source defines a network as "a form of associative behavior among firms that helps expand their markets, increase their value-added or productivity, [and] stimulate learning [to] improve their long-term market position" (Bosworth and Rosenfeld 1993, p. 19). Firms in a given network may or may not serve similar markets or be members of the same product chain. Bosworth and Rosenfeld describe two basic types of networks. *Vertical* networks consist of firms at different stages of the production chain or in the same markets that form an association to engage in joint marketing or share information regarding production or product development. *Horizontal* networks are made up of firms that share similar technology or service needs, whether or not they are in the same product chain.[8]

In practice, the organizational structure of the networks varies significantly, as does the degree of formality involved. Networks are most common in Europe, particularly in Italy and Denmark, but their numbers are growing in the United States. In 1994, the U.S. Department of Defense, through its Technology Investment Project, established a three-year networking program called USNet, a consortium of manufacturing assistance organizations in eleven states. USNet is designed to identify the specific forms of interfirm collaboration that are most effective in encouraging manufacturing modernization. At the local level, a growing number of cities and states are establishing their own networks to encourage information exchange, joint problem-solving, and innovation.

III. Elaboration

Any production technology involves the direct, physical combination of materials, labor power, and equipment to manufacture goods, as well as the coordination of relationships. These relationships may be internal to the firm (such as between different functions, for example, production and R&D) or they may extend outside the firm (relationships with suppliers of intermediate goods and capital equipment, or with customers or competitors). Because different production technologies combine factors and manage relationships differently, they also have different implications for geography. Essentially, the debate over Fordist versus flexible manufacturing constitutes a debate over trends in industrial organization and their impact on the development prospects of particular regions. The debate is carried on at a fairly abstract level, but clearly it is a simplification to reduce all production technology to two general paradigms. Nevertheless, the flexible-specialization approach has succeeded in making industrial organization a central concern in economic development. Local developers benefit from thinking not only about the basic locational needs of industry (quality labor, infrastructure, regulation, and taxes), but also about ways local firms can increase innovativeness, quality, and flexibility through adoption of advanced process technologies, their R&D activities, and the ways in which they choose to manage linkages with suppliers and customers.

The New Flexibility: Machines, Firms, Economies

Early writing on flexible production focused on broad shifts in technology—that is, the question of whether the most competitive businesses are, in fact, replacing the standardized production-oriented technologies with new, more flexible technologies. But what does "flexible" mean? Gertler (1988) identified a "hierarchy of flexibility stretching from the individual machine on the shop (or office) floor to the very basis of organization within the economy and the society in which it is embedded" (p. 420).[9]

At the first level of the hierarchy is flexibility in production machinery. New types of capital equipment that take advantage of computer technology allow producers to reprogram rather than completely retool as they introduce qualitative changes in products or even wholly new products. The result is that plants are becoming obsolescent at a slower rate. Perhaps more important, it is becoming easier for companies to introduce variations in product to meet niche markets, a type of customized mass production. In some cases, this means that the time from plant setup for a new product to actual production is reduced; in other cases, plants are able to vary products significantly in real time. Not surprisingly, the study of the rates at which firms are adopting new technologies has become a small industry. The federal government (U.S. Department of Commerce 1989, 1993) as well as numerous state governments have administered surveys designed to gauge the degree to which producers are incorporating these new technologies. The data inform manufacturing extension programs and other modernization strategies designed to help diffuse the new technologies throughout the economy.[10]

Individual pieces of flexible machinery do not operate in a vacuum. The second level of the "flexibility hierarchy" involves combining the machinery and business functions into an integrated system (often called computer-integrated manufacturing—CIM). Computers can be used to help design, draw, and test parts (computer-aided design—CAD, computer-aided engineering—CAE), the instructions from which are then used to control the machines that actually produce the product (computer-aided manufacturing—CAM). This process may be aided by robots and automated material-handling systems, as well as a computer network within the factory (local area network—LAN) to different departments such as procurement, design, and production. Flexibility in production is also improved through relationships with outside suppliers and customers. It is through the linkages between firms that spatial implications become manifest. Firms may use just-in-time (JIT) scheduling and delivery systems to reduce inventory costs and interfirm computer networks to manage this process through the electronic exchange of data related to design specifications and delivery orders.

The third level of the flexibility hierarchy is of the greatest interest to theorists of flexible specialization. This is the level of strategy and management. The types of contracts negotiated between buyers and suppliers are important here (the management of transactions); they in turn are tied to the general strategy on which a given firm chooses to compete. Flexible-specialization theorists such as Piore and Sabel (1984) argue that demand conditions are such that firms must increasingly specialize in niche markets; the days of mass markets of consumers with undifferentiated tastes are coming to a close. Firms can serve niche markets better by working closely with suppliers—perhaps even outsourcing a greater share of the production of a given good—to produce a greater variety of goods in smaller quantities. The need for continuous innovation, emphasized in studies of Japanese business practices, also dictates coordination with suppliers. Successful implementation of this strategy may require longer-term, stable relationships with business partners rather than the arm's-length, least-cost type of transactions presumably common in large-scale manufacturing.[11] The strategy suggests that smaller firms will be able to compete with larger firms more effectively than ever before. The greater the degree to which firms outsource production, the more apparent is the fact that one firm's competitiveness depends crucially on the competitiveness of its partners.

Researchers have found that crucial to the success of flexibly specialized firms are cultural and social norms that foster trust between contracting parties.[12] In other words, the means by which contracts—generally conceived as agreements between parties—are governed is important. For instance, trust-based transactions are particularly apparent in several smaller regional economies of northern Italy (called "Third Italy"). Long-standing familial ties, a close meshing of business and social relationships (it is preferable to do business with family and friends), and the sense of community that these dynamics engender are common in these regions. Due to a recent track record of strong industrial performance, the economic structure of Third Italy has become a model that development analysts have attempted to export to other places. One difficulty, however, is that the economic and social structures go hand in hand in determining the success of flexibly specialized economies.[13] Exporting a social structure—assuming it can be operationally defined—is obviously a near impossible task, even if it is considered desirable.

Flexibility and Agglomeration

Two perspectives from which the influence of flexible production on spatial development patterns has been examined are the transaction-cost approach of A. J. Scott and the new industrial districts (NIDs) literature that grew out of the original studies of Third Italy. The latter takes Marshall's (1961 [1890]) analysis of the source of dynamic external economies in dense urban complexes as a point of departure. Dynamic external economies are the beneficial effects on a given firm of the general growth in industry (Feser 1998a). The benefit may be manifested in reduced costs, enhanced productivity, or superior innovativeness. One important source of these economies is the knowledge and expertise shared by firms in an industry (so-called "knowledge spillovers"); a firm benefits (or learns) by interacting with industry leaders just as the professional developer learns by networking informally with peers. As the industry expands, so does the common pool of knowledge, thus engendering further growth.

Scott's approach is also based on the concept of external economies, though with modifications based on the theory of transactions costs. As alluded to above, one can think of a firm as composed of several distinct functions—production, administration, management, research and development, and so on (Robinson 1931). The activities of the typical firm may be further subdivided within each of the broad functional areas. For example, some firms may produce in house a large share of the manufactured inputs they need. When this is the case, the firm effectively produces multiple products, some for final markets and some for intermediate markets (in which the only customer may be itself). But the cost-minimizing scale (level of output) of production of each good may vary significantly. For example, it may cost a firm less to subcontract duplicating and printing services to outside companies than to maintain its own printing office. Print shops, by serving a larger customer base, are able to attain economies of scale the individual firm cannot achieve. The same may be true of certain manufactured inputs or other business functions.

In an economy or region in a relative state of underdevelopment, the typical firm may have to produce most of its own inputs and handle most of its basic functions in-house. The market may not be sufficient to support other companies focused exclusively on duplicating and printing, booking travel, managing secretarial tasks, conducting R&D, or producing intermediate inputs of various kinds. This is a problem typical of underdeveloped economies in many parts of the world. However, as the market expands with economic growth, the establishment and survival of specialized firms become viable; therefore, the original firm gains the option of spinning off many of its internal activities—that is, if it is cost-effective to do so. Adam Smith described the advantages of division of labor internal to the enterprise in terms of a pin factory, where the firm reduces costs by specializing tasks (one worker makes the pin shaft, another the head, a third joins the two, and so on.). But, with the growth of the market, the internal division of labor may be converted to an external one, and tasks may be spread among multiple firms. This idea was originally outlined in rigorous fashion by Stigler (1951), based on Young's (1928) discussion of the link between the division of labor and economic growth.

To the degree that firms in a production chain benefit from locating in proximity to each other, these ideas have clear implications for regional growth. As a metropolitan area grows, the possibilities for a finer division of labor are improved. Enterprises in the large, growing regions enjoy advantages vis-à-vis those in smaller or rural areas. In this way, industry in large, growing places is able to increase its competitive edge, thus spurring additional growth. This is, in effect, a type of cumulative causation argument consistent with spatial polarization models laid out in chapter 5, post-Keynesian and new endogenous growth theories in chapter 6, and new trade theory in chapter 7. Of course, an important assumption embedded in the model is that firms are close to their suppliers; otherwise, the growth of the local economy in which a given enterprise is situated is irrelevant, since that enterprise could secure what it needs from other places (for example, New York, Los Angeles, London, Tokyo, or Hong Kong). Thus, the question of proximity between firms and their corporate partners is critical to the question of what flexible specialization means for local economies.

In his study of flexible production, Scott (1986, 1988, 1992) meshes the basic framework of Stigler (1951) with the modern theory of transactions costs (Coase 1937; Williamson 1975, 1985). Transactions costs theory focuses on the factors that determine whether a firm chooses to produce a certain good itself or, instead, contracts with another firm to produce the good. Businesses have a choice of organizing production through the market (for example, outsourcing inputs) or internal hierarchies (establishing its own division for the manufacture of the input). The firm's choice depends on the costs of the transactions associated with each alternative. In conditions where there is a high degree of uncertainty in terms of the type, quantity, or quality of inputs required, firms might choose to produce that good internally. Alternatively, when intermediate inputs are standardized, the firm may find it cost-effective to use the market (other firms) as its source. The study of transactions costs is the study of the conditions under which one alternative is better than the other. Clearly, the superstructure of laws, regulations, and social norms that govern market activities are a critical element of their study. Firms may choose one course of action in one country that would be extremely costly in another. To a more limited degree, this may also be the case across regions in the same country. Although we have cast this discussion in terms of manufactured inputs, it may also be applied to other types of factor requirements.

Scott shows how these ideas apply in the context of flexible specialization. If firms increasingly are using flexible rather than standardized methods to meet customer demands, and this flexibility requires relationships with nearby partners, then we should observe an increase in the incidence of local linkages and a re-agglomeration of many industries. Why should firms seek flexible methods rather than those of standardized production? Because as the degree of global competition increases, it becomes more difficult for firms to predict demand. In addition, consumers in industrialized economies are more savvy than ever, and they seek more specialized and higher-quality goods than the typical consumer at the start of the Industrial Revolution and even during the boom period that immediately followed World War II. Firms achieve the necessary flexibility by spinning off in-house functions and relying to a greater degree on contract suppliers. But why should firms locate near their business partners? Because, in order to maintain a high standard of quality and a continuous stream of innovations, businesses must work more closely with their partners than simply placing orders for given quantities of goods, delivered at stated times, with agreed-upon rates of defectiveness. The benefits of face-to-face contact possible when firms are located in proximity are important.

One implication of these dynamics is that flexible-production methods and techniques will likely lead to a re-agglomeration of economic activity. This essentially empirical question is a major point of contention in the regional development literature. At issue is to what degree the forces encouraging agglomeration by firms (for example, the importance of proximity in ensuring flexibility of production between linked firms, face-to-face interaction, and so on) outweigh sweeping general improvements in technology and infrastructure that would seem to encourage (or at least permit) greater spatial dispersion (such as advances in telecommunications technologies). A specific issue that has inspired considerable debate is whether just-in-time inventory and delivery systems will necessarily lead to tighter co-location of suppliers and their customer firms. Unlike many specific flexible-production technologies, the potential relationship between JIT and spatial development patterns is fairly easy to identify (though, as history has shown, certainly not to prove).

JIT attempts to improve the quality of final goods through the use of smaller-batch runs that permit quicker identification of problems in the production process. Not only are defects and errors detected more rapidly, but their costly and disruptive consequences are limited. The system also minimizes inventory, as suppliers ship goods to the end-market producer with greater frequency (rather than intermittently) and in large quantities. The objective is to approximate, to the extent possible, a continuous manufacturing process throughout the entire production chain. As noted by McCann and Fingleton, "it is necessary for the customer firm to have individual shipments of goods delivered in exactly the size and frequency it requires, otherwise its internal production operations may become hampered" (1996, p. 494).

The nature of JIT suggests that coordination and transportation costs may be significant enough to necessitate closer proximity between firms that adopt the system. And, if such a technique became widespread in the manufacturing sector, we might expect to observe a general clustering of industry in space. Testing these propositions is by no means easy, however, and most analyses have been anecdotal. An exception is the study by McCann and Fingleton (1996), which found tighter spatial linkages following JIT adoption among a small sample of Scottish electronics firms. What distinguishes their analysis from many others is that it involved an attempt to develop a theoretical model of JIT-based production in space that could be subjected to rigorous empirical testing. Many flexible-production ideas require this kind of explicit model-building before they can be properly evaluated.

As a final example of how technology and organization bear on the performance of regional economies, consider Saxenian's (1994) comparison of the high-technology regions of Route 128 near Boston and Silicon Valley in California's San Francisco Bay Area. The former is described as a region dominated by vertically integrated, self-sufficient companies concerned with maintaining proprietary control over innovations and technology. This control orientation is manifested even in the spatial dimension; that is, companies along Route 128 are located on large, self-contained, and campus-like tracts of property. There are few informal institutions where workers from different companies can mingle and exchange ideas. In contrast, the history of Silicon Valley is the history of the independent, garage-based entrepreneur. Even today, after these entrepreneurs have developed their own large companies and associated facilities, the business environment remains more informal and workers move between employers often, taking their accumulated knowledge and ideas with them. Again, the spatial environment plays a role. The sprawling development patterns of the South Bay Area encourage the high degree of social networking that has become the region's hallmark. Saxenian concludes that:

> Silicon Valley continues to reinvent itself as its specialized producers learn collectively and adjust to one another's needs through shifting patterns of competition and collaboration. The separate and self-sufficient organizational structures of Route 128, in contrast, hinder adaptation by isolating the process of technological change within corporate boundaries. (1994, p. 161)

For the local economic developer, flexible-production theories provide a hypothesis of what determines business competitiveness in the increasingly global economy. Therefore, they provide a framework for evaluating the likely performance of local enterprises. At the same time, however, they implicate an entirely different set of locational characteristics as critical in shaping the aggregate performance of regions. Industrial structure and organization, the degree to which firms exchange information through formal and informal channels, and even the spatial structure of the broader urban environment—all of these call for attention, along with the more typical concerns of infrastructure, workforce quality, regulations, and investment capital.

New Industrial Districts

The set of locational characteristics implied by flexible-production ideas is the focus of the new industrial districts (NID) literature. The concept of the industrial district originated with Alfred Marshall's *Principles of Economics* (1961), in which Marshall described the advantages smaller firms derive from locating in spatial proximity—that is, in dense industrial complexes. The advantages are a function of dynamic external economies; as the local economy grows, so does the availability of inputs, the pool of skilled labor, and the volume of knowledge spillovers. All this is noted, in one form or another, in the discussion above. Rabellotti (1995) characterizes the industrial district as a spatially concentrated cluster of sectorally specialized firms, with a strong set of forward and backward linkages, a common "cultural and social background linking economic agents and creating a behavioral code, sometimes explicit but often implicit," and a network of public and private supporting institutions (p. 31). The NID literature has focused foremost on examining particular dynamic regions for evidence of these characteristics. Although Third Italy has received most of the attention, the number of case studies of other regions and countries has steadily broadened. Now one can find studies not only of the usual suspects in the United States (such as Silicon Valley), but also regions in Asia and the developing world (Mexico, India, and Brazil, to name a few).[14] These cases play a strong role in motivating economic development strategies related to industry clusters in the United States and Europe (Rosenfeld 1995).

NIDs have an explicitly normative dimension. In fact, industrial districts have effectively been defined as places where the dominant industries employ flexible-production methods *and* are highly competitive. In other words, flexible production is, by definition, competitive production. On the methodological side, what this means is that few researchers have conducted studies that identify specific characteristics of a flexibly specialized economy and then examine their incidence across a large cross-section of areas. Presumably, such an approach might turn up places that seem to follow the pattern of flexible production, yet, in a performance sense, are struggling. Obviously, one factor in limiting these kinds of studies is lack of good data on many of the contextual factors emphasized in the flexible-production literature. Nevertheless, one can also gain the impression—which is dubious at best, given available information—that the type of industrial structure and organization highlighted in the flexible specialization literature is a sure route to sustained economic growth.

On the conceptual side, it has become clear that a normative basis of the NID perspective can lead to an excessively limited picture of what determines regional performance. What about those successful regions that embody many of the characteristics of older, presumably discredited development models? According to Gray, Golob, and Markusen:

> Among the stronger, more resilient regions in a country like the United States, however, there are many whose industrial structures and cultures cannot be considered flexibly specialized. These have been understudied and underappreciated as alternative development models. (1996, p. 652)

Markusen (1996) maintains a normative view of industrial districts, identifying them generally as "sticky places." Sticky places are regions, in this era of the increased mobility of business, that are able to both attract and keep industrial activity over time. She then develops a typology of industrial districts: (1) the Marshallian; (2) the Italianate variant to the Marshallian; (3) the hub-and-spoke; (4) the satellite industrial platform; and (5) the state-anchored industrial district. The core features of each type of industrial district are summarized in table 10.1. Essentially, they describe the most common types of regional industrial structures. Markusen found evidence of each type after examining a subset of dynamic metropolitan economies in the United States, Japan, Korea, and Brazil.

Markusen's typology implies that flexible specialization is not the "new orthodoxy." That is, many of the characteristics of the temporarily discredited Fordist model are also associated with highly dynamic places. For example, the hub-and-spoke district is dominated by one or two large, vertically integrated manufacturers (such as Boeing and Microsoft in Seattle), which, arguably, enjoy significant internal economies of scale. These firms have not spun off much of their production, even as their region (indeed, the world economy) has grown. Clearly, advantages remain that are associated with large scale. If the world is not adopting the flexible-specialization model wholesale, what is its relevance for development practice? Perhaps the most important lesson for the local economic developer is that industrial organization matters, although the type of organization that is most successful in particular places may vary. When designing ways to help local industry reduce costs, increase innovation, and become more productive, developers may find that the most appropriate ways to accomplish these goals are through initiatives that influence business structure and organization.

Flexible Production and Economic Change

Like many ideas of economic development, early ideas about flexible production were received with a degree of enthusiasm and optimism that at times seemed to be messianic. Enterprise flexibility, particularly as it is

TABLE 10.1

A Typology of Five Industrial Districts

District type	Characteristics
Marshallian *Examples: None given*	Business structure dominated by small, locally owned firms Firms enjoy few internal scale economies Substantial local interindustry trade Long-term buyer-supplier contracts and business partnerships Few linkages with firms outside the region
Italianate *Examples: Third Italy*	Characteristics of Marshallian plus: Exchanges of personnel between buyers and suppliers common Cooperation between competitors to share information, risk Strong R&D, design function Strong role for government in regulation and boosterism
Hub-and-spoke *Examples: Seattle; central New Jersey; Toyota City, Japan*	Business structure dominated by one or several vertically integrated firms Dominant firms' focus is on global, not local, community Investment decisions, with global effects, made locally Dominant firms maintain strong external rather than internal linkages Internal economies of scale important
Satellite industrial platforms *Examples: Research Triangle, North Carolina; Kumamoto, Japan*	Business structure dominated by large branch plants Internal scale economies important Minimal local interindustry trade Key investment decisions made elsewhere Labor market external to the region
State-anchored districts *Examples: Santa Fe; San Diego; Madison, Wisconsin*	Business structure dominated by public-sector institution(s) Local firms serve as suppliers to dominant institution(s) Low rates of turnover among local businesses Short-term contracts between institutions and suppliers Disproportionate shares of clerical and professional workers

Source: Adapted from Markusen (1996), table 1. See source for full list of characteristics.

manifested in the industrial district of prototypical Third Italy, appeared to suggest, to many researchers and local officials concerned with local development questions, a new and promising alternative route to regional competitiveness. Piore and Sabel's (1984) thesis of a potential new era of flexibly specialized production raised the possibility that viable communities might be organized around small- and medium-size firms that cooperate as much as they compete, that have long-standing ties to their localities, and that continuously innovate as a means of competing with larger firms. The model implies that local economies need not be subject to the whims of large, impersonal, vertically integrated multinational companies—that is, that local workers could have a greater hand in determining their own economic fate. According to Bellini,

> Third Italy was ideologically fascinating because it seemed to give irrefutable evidence of the possibility of alternative, socially progressive paths of capitalist growth and widened the scope of public policies, giving full intellectual and political dignity to a number of micro-interventions at the territorial level. (1996, p. 3)

However, as the literature deepened, potential negatives associated with industrial flexibility became increasingly evident. Indeed, it became clear that there are different versions of flexibility with fewer perceived benefits for local control. In some cases, these variants evolved in the very regional economies that early studies cited as benchmark examples of industrial districts. In a major critique that included exhaustive treatments of the small-firm debate, industrial districts, and interfirm networking, Harrison (1994) uses case examples to illustrate the adverse effect corporate flexibility can have on workers, their earnings, and their communities. In particular, he describes how regions that once seemed to characterize the flexible-production model have evolved and changed as global competition has intensified. He found evidence that the use of child labor, the exploitation of immigrants, and poor wages and employee benefits were common practices among some small- and medium-size enterprises in Third Italy, which increasingly were being squeezed by Third World competitors. In the intense competitive environment, new loci of corporate control were emerging in regions formally dominated by independent, small manufacturers. Harrison demonstrates that mergers, acquisitions, and vertical integration may provide corporate flexibility in some forms, although with different implications for regional economies.

These findings suggest that the flexibly specialized economy composed of small independent producers does not necessarily constitute a permanent development model. Indeed, flexible specialization may easily represent only one stage in the development path of a given region. Small firm-oriented flexible production must be viewed in a dynamic context, rather than as an assured new model of competitive success (Asheim 1996).

Summary

Flexible-production theories are consistent with a general trend in regional development thought toward (or back to?) a focus on industrial organization and structure, interindustry linkages, and externalities and agglomeration economies. Theories of concentration and diffusion (growth poles, unbalanced growth, cumulative causation, core-periphery), new growth and trade theory, and entrepreneurship and flexible-production theories—all, to one degree or another, focus on the way work is organized and managed between and within firms. This organization, in turn, has implications for spatial development patterns, generally through its influence on spatial externalities (spillovers) and innovation. This is a stark contrast to the theories of economic base, neoclassical growth, and neoclassical trade. In economic base theory, the focus is on macroeconomic aggregates (basic versus nonbasic employment), whereas neoclassical theories analyze the workings of the price mechanism in a world of constant returns and atomistic producers. Flexible-production theories imply neither an export-focused nor a hands-off approach, but one that attempts to build the capacity of firms in a given region to continuously innovate, learn, and adapt to rapidly changing economic circumstances.

Discussion Questions

1. Flexible-production technologies may include production equipment, methods of organizing production, and worker-management strategies. Development strategies aimed at increasing firms' adoption of advanced production equipment are common. But should local developers also attempt to improve the adoption of flexible workforce-management techniques? How might such a strategy be pursued?
2. Identify several benefits available to enterprises through formal interfirm networks.
3. Are there any ways in which the local political context might influence the means by which regional firms organize production? Is the national political context likely to be the biggest influence on production organization?

4. How does the dominant industrial culture of an area (for example, steel production in Pennsylvania, textiles in North Carolina, computers in the San Francisco Bay Area) influence the competitiveness of local producers in the dominant sectors? In what ways might the influence be different for firms outside the dominant industries?
5. The adoption of flexible-production technologies does not ensure that firms will continuously innovate and develop new products. Are there ways local developers can influence the propensity of firms to innovate?
6. How do the predictions of flexible-production theories regarding the spatial distribution of economic activity differ from those offered by product-cycle theory?
7. Is there such a thing as cooperative competition? That is, how might highly competitive firms also seek ways to collaborate?

Notes

1. The question of whether flexible production is actually replacing so-called Fordism as a dominant mode of organizing production generated a spirited debate (note that early adherents to Piore and Sabel's thesis were actually bolder in their predictions about this than were the original authors themselves). See Holmes (1986), Scott and Storper (1987), Schoenberger (1988), Gertler (1988, 1993), Pollert (1988), Amin and Robins (1990, 1991), Martin (1990), Sayer (1990), and Wilson (1990).

2. In addition to chapter 9, see Ettlinger (1997) for a discussion of the debate surrounding the importance and implications of small firms in regional economies.

3. For example, some flexible manufacturing technologies such as just-in-time (JIT) sourcing imply contracting practices that tie suppliers closely to their customers. Under JIT, large assemblers may opt to maintain longer-term relationships with a few key suppliers rather than purchase components from a larger number of competing firms. The small group of select suppliers are favored, but they are also dependent if they are discouraged from selling to multiple producers (as research has shown is sometimes the case). This may have important implications for regions with particular industry and firm structures. A region dominated by small- and medium-size firms may, from a cursory point of view, appear to fit the profile of the Marshallian industrial district. But if those firms are suppliers serving nonlocal firms, the flexible-production model of continuous innovation, networking, and information sharing may mean less for explaining regional fortunes than the decisions of one or a few large final-market (and often multinational) companies.

4. Technology-adoption benchmarking surveys (or technology-use assessments) have become a common practice in economic development. For examples, see NC ACTS (1996), Rephann and Shapira (1994), Swamidass (1994), Youtie and Shapira (1995), and Bergman, Feser, and Scharer (1995).

5. See Appendix 4.1 for additional citations of specific applications of cluster policy at state and local levels.

6. Clusters typically are defined as establishments in related industries that are linked in some way (for example, through buyer–supplier relationships) *and* that tend to co-locate in geographic space. The latter attribute is assumed in the flexible-production literature. However, earlier studies of interindustry relations distinguished industrial clusters from industrial complexes. As discussed in Appendix 4.1, an industrial complex consists of industries in an industrial cluster that are in the same location (co-located). In a test of spatial and economic clustering, Feser and Sweeney (1998) found evidence that linked firms (particularly those in technology-oriented sectors) actually do have a greater propensity to cluster geographically. Sweeney and Feser (1998) tested whether small, single establishments have an "above average" tendency to co-locate, a key postulate of many flexible-production theories. That study found evidence of spatial clustering by establishments employing between ten and fifty workers; no spatial clustering was found for the smallest or larger (more than fifty workers) plants.

7. See Boekholt (1997) and Roelandt et al. (1997) for alternative typologies of cluster concepts and policies.

8. Boswell and Rosenfeld (1993) define a third type of network as a knowledge network, made up of firms with few commonalities in terms of product chain or market, that band together to share information regarding business practices. These firms "meet to identify and solve common problems, exchange information, and stimulate continuous learning and improvement." The knowledge network is effectively a special type of the horizontal network.

9. It should be noted that, although much of the flexible-production literature focuses on small firms and often customized producers, large, mass producers may also adopt flexible-manufacturing techniques. In fact, some evidence suggests that the highest rates of adoption of flexible techniques and practices are among larger producers (Bergman and Feser 1998).

10. Studies of the determinants of technology adoption include Benvignati (1982), Wozniak (1987), Antonelli (1990), Dunne (1994), Julien and Raymond (1994), Little and Triest (1996), and Harrison, Kelley, and Gant (1996a, 1996b).

11. Although this chapter concentrates on flexible-production ideas examined specifically in the regional development literature, corporate flexibility is the subject of a large literature in strategic management. For example, see Helper (1991a, 1991b, 1994) for analysis of buyer–supplier practices in the U.S. automotive industry. Trends in industrial organization and outsourcing, firm "core competences," and lean manufacturing are also addressed in Waitt (1993), Klier (1994), Foss and Knudsen (1995).

12. This is often described as the concept of "embeddedness" (see Granovetter 1985, Harrison 1992, Asheim 1996).

13. Ettlinger (1994) provides a comparison of the role of differing national contexts (mode of production, social relationships) in influencing the development paths of various industrialized countries. She argues (pp. 161–162):

> Critical ingredients of long-term localized development, regarding both productivity and social welfare, include a production system and set of social relations suited to partnership principles of cooperation and collaboration, appropriate organization of local interests to achieve consensus, and finally,

an active local government that articulates the needs of workers and firms through both supply-side policy (education, training, service provision) and indirect relations with the private sector through incentives that influence competitive firm behavior amid competing production systems in the global economy.

14. See also Harrison (1992) and Park (1997).

11

Economic Growth versus Economic Development

The theories discussed in this book describe the economic development process quite differently. Some are concerned primarily with the near-term expansion of the local economy. These theories may be classified as theories of economic growth. Economic base theory, neoclassical growth theory, and interregional trade theory are essentially economic growth theories; other theories deal with evolutionary and structural change occurring over a long time horizon. Attempts to explain economic development as a long-term process may be treated as economic development theories. Staple theory, sector theory, growth pole theory, and entrepreneurship and flexible-production theories consider economic structure in detail, rather than focusing on one or two regional sectors; they also focus on structural change occurring over a long time. Product-cycle theory has one foot in each camp. As an extension of trade and location theory, it represents a contribution to growth theory. When used to examine the innovation process or the organizational structure and technology of firms, product-cycle theory becomes development theory.[1]

The distinctions between economic growth and economic development are introduced in chapter 2, where Robert Flammang's ideas are presented. In this chapter, we review Wilbur Thompson's work as a synthesis of the growth and development process. His insights have much to offer economic developers. The practice of economic development in the United States is, in fact, concerned primarily with achieving economic growth at particular locations. As a result, economic development, as a process, is often reduced to and equated with economic growth. An understanding of the distinctions between growth and development, accompanied by the application of this understanding in the field, offers an enormous opportunity to expand economic development practice in many useful directions.

I. Thompson's Synthesis

Wilbur Thompson (1965) brings together concepts of economic base, linkages, product cycle, and entrepreneurship, while introducing the importance of organization (oligopoly and unionization) in the development process. Although his model is demand-driven and distribution-sensitive, he recognizes the key factor on the supply side: entrepreneurship. Thompson argues that economic base theory (and the economic base model) provide a useful way to understand the local economy if we are concerned with near-term economic growth. Consistent with Keynesian short-run analysis, the theory is especially useful for estimating business cycle impacts. In the near term, the industry mix, particularly the exporting industries, determine the rate of growth. They influence the entire local economy through linkages that transmit multiplier effects.

Economic developers, city planners, and others concerned with long-term change must look elsewhere for appropriate theory. For us to understand the local development process, we must examine the competitiveness of the local economy over decades and compare it to other areas.[2] In the long term, urban growth leads to larger markets and more interaction in industrial complexes, both of which attract producers who can realize economies of scale. Furthermore, larger markets attract more sophisticated business and professional services and require additional public services. Urban institutions facilitate adaption to external forces and support locally initiated change. As the size of the area and period of time increase, local activity comes to the fore as the prime mover and catalyst for development. Here, Thompson echoes Blumenfeld's (1955) thesis about the importance of city-building activities (see chapter 3).

Development supports the sequence from invention to innovation and then to commercialization. It leads to higher levels of welfare in terms of (1) income level, (2) income distribution, and (3) income stability. Thus, Thompson defines development in terms of welfare economics, rather than as jobs and investment (tax base), the pillars of economic development practice. Income *level* is a function of the skill and power of local organizations; *skill* refers to the ability of local producers to create and successfully market new or income-elastic products. *Power* refers to oligopolies and labor unions that use size and collusion to garner market share. High wages in the oligopoly sector "roll out" in order to increase hourly earnings throughout the local labor market.

As for income *distribution*, economic growth occurs in localities where the industries export new products or income-elastic products. Greater demand begets greater supply, which requires higher workforce participation rates and puts upward pressure on wages and other costs. As the wage roll-out effect continues to operate, demand growth reduces income inequality by employing marginal workers. Gradually, the area may attract low-income in-migrants seeking job opportunities, with the result that the local level of unemployment may increase.

Income *stability* depends on several factors. Places that specialize in producer durables tend to be most unstable because investment expenditures are less stable than consumption expenditures. Places with older establishments are more susceptible to cutbacks and closings. City size is associated with stability because, generally, the economic base of cities tends to become more diverse, and therefore more stable, as size increases.

Large places are in the best position to enjoy stable growth over time because their export base contains a mix of new and income-elastic products as well as mature and inelastic ones ("breadth"). Size also bestows the ability to support innovation, which is needed to change industry mix for long-term viability ("depth"). Thompson offers an insightful and invaluable statement about what makes a local economy competitive:

> The economic base of the larger metropolitan area is, then, the creativity of its universities and research parks, the sophistication of its engineering firms and financial institutions, the persuasiveness of its public relations and advertising agencies, the flexibility of its transportation networks and utility systems, and all the other dimensions of infrastructure that facilitate the quick and orderly transfer from old dying bases to new growing ones. A diversified set of current exports—"breadth"—softens the shock of exogenous change, while a rich infrastructure—"depth"—facilitates the adjustment to change by providing the socioeconomic institutions and physical facilities needed to initiate new enterprises, transfer capital from old to new forms, and retrain labor.[3] (1968, p. 53)

With respect to the factors of production, Thompson argues that larger urban areas not only have advantages in labor force and immobile capital but in entrepreneurship as well. Original natural resource endowments that gave rise to the city at that location are continually modified by the growing built environment that satisfies workers and, more important, attracts entrepreneurs. These compelling advantages led Thompson to hypothesize the "urban size rachet" effect that results in sustained advantages for larger cities (1965, pp. 21–24). Cities are the locations where entrepreneurship and immobile capital combine to achieve growth.[4]

Using concepts from shift-share analysis and industry (not product) life cycles, Thompson presents the relative importance of industry mix and regional share in large versus small cities. Large metropolitan centers spin off industries to smaller peripheral areas. Large cities can grow above the national average by capturing new products while spinning off routine work. Smaller cities are more volatile, winning or losing routinized production. The former become high-wage, high-skill places where labor unions are more prominent; the latter are low-wage, low-skill, unorganized towns. He articulates the filtering-down dynamic in the following way:

> The larger, more sophisticated urban economies can continue to earn high wage rates only by continually performing the more difficult work. Consequently, they must always be prepared to pick up new work in the early stages of the learning curve—inventing, innovating, rationalizing, and then spinning off the work when it becomes routine. In its early stages an industry also generates high local incomes by establishing an early lead on competition. The quasi-rents of an early lead are in part lost to the local economy, as dividends to widely dispersed stockholders, but in part retained as high wage rates, especially if strong unions can exploit the temporarily high ability to pay. It would seem, then, that the larger industrial centers as well as the smaller areas must run to stand still (at the national average growth rate); but the larger areas do run for higher stakes.
>
> In order to develop, it seems that the smaller, less favored urban area must attract each successive industry a little earlier in the industry's life cycle, while it still has substantial job-forming potential and, more important, while higher-skill work is required. Only by upgrading the labor force on the job and generating the higher incomes—hence the fiscal capacity—needed to finance better schools, can the area hope to break out of its underdevelopment trap. By moving up the learning curve to greater challenge and down the growth curve toward higher growth rates for a given industry, an area can encourage the tight and demanding type of local labor market that will keep the better young adults home, lure good new ones in, and upgrade the less able ones. (Thompson 1968, pp. 56–57)

Thompson recognizes that the overall impacts of these processes are not easily determined. Positive and negative feedback represent countervailing effects on local growth and change. Equilibrium and disequilibrium forces are reflected in the struggle between market power and overpriced labor. Positive versus negative outcomes are partly the result of timing.

The Occupational–Functional Dimension

Twenty years later, Wilbur Thompson and his son, Phillip (1987), extended these ideas to incorporate an "occupational-functional" dimension. They argue that industrial targeting—the recruitment of large firms in attractive industries that are feasible, given local resources—needs to be broadened. Specifically, it is too crude an approach unless coupled with an occupational-functional focus that defines an area's "appropriate work." The functional specialization that shapes the local industry mix reveals what the place makes—for example, textiles or automobiles. The occupational-functional dimension identifies what the place does, that is, the type of work that is most prevalent. Thus, a factory town offers production and assembly work, whereas a college town, state capital, or headquarter city offer information-intensive office work.

Five occupational specializations representing different types of work suggest ways to achieve competitive advantage: routine production, precision production, research and development (R&D), headquarters administration, and entrepreneurship.[5] One locality may have strength in more than one area, but rarely in more than two or three areas. The occupational-functional dimension can be used to indicate where, among the cities in the urban system, new work (that is, product innovation) can best be done and where existing work can be done most efficiently.[6]

Development occurs as localities gradually change skill levels and occupation mix, which in turn change functional specializations and, gradually, the basis of competitiveness. In the near term, however, the obvious strategy is to leverage existing occupational strengths, be they routine production, precision production, R&D, central administration, or entrepreneurship. At headquarters locations, dominant companies may promote innovation through internal investments in new products or through skillful acquisition of patents and young growth companies.

With creative and efficient firms and individuals, the local economy can change and add specializations and continue to enjoy a viable export sector. The local-basic dichotomy from economic base theory should be made a trichotomy, in order to understand more fully the role of the nonbasic sector. Thompson distinguishes developmental services from routine distributive services, in addition to basic industries. In the near term, the city's economic base is its export manufacturing. In the long term, its economic base depends on the availability and quality of its developmental services, services that include a sound educational system, diverse health services, rich cultural activities, and a range of sophisticated business and professional services that can sustain the creation of new products and new business enterprises.[7]

In conclusion, Thompson's contributions offer a useful synthesis for understanding local economic development. Economic growth can be assessed by analyzing near-term changes in the urban area's industry mix. Growth typically leads to more employment and aggregate income, a larger tax base, and higher property values. The more complex process of economic development can be examined by gauging the structural change that occurs over the long term. Localities experiencing economic development over time should be able to sustain economic growth that leads to improvements in per capita income levels and to less unemployment and instability. Although economic growth and economic development are related processes, development is more fundamental. Economic development leads to and sustains competitiveness; economic growth results from competitiveness.

II. Strategic Applications

Local economic developers interested in applying ideas about economic development may face resistance. First, local public officials and business and community leaders rarely view economic development as a long-term process requiring steady guidance over time. In many places, economic developers are directed to determine a quick fix to high unemployment or a shrinking tax base. Results are expected before the next election, which may occur within two or three years.[8] In the worst case, hiring an economic developer is considered the "affordable" solution—it distracts attention from more profound local problems such as poor public schools, inadequate physical infrastructure (roads, water, sewers, utilities), outdated systems of taxation or regulation, contentious race relations, or weak local leadership. Economic developers will find little or no support for developmental strategies in these places.

Second, the goals articulated by Thompson are not mainstream in economic development practice. It is rare to find a jurisdiction committed primarily to increasing per capita income, reducing income inequality, and lessening income instability. Most jurisdictions designate as their primary development goals jobs and investment (tax base). Localities, however, seem increasingly willing to embrace the goal of greater economic diversity, presumably as a way to experience greater stability. Furthermore, income inequality is an issue that is increasingly hard to ignore in economic development circles.

Developers encouraged to formulate an economic development strategy consistent with Thompson's thinking should begin with an analysis of the global economy and the urban system. The former highlights the economic location of the local labor market area (see chapter 2). The latter helps identify the linkages connecting the area to other labor market areas (see Appendix 4.1). With these insights, developers should (1) examine both the evolution of the locality's industry mix and its occupational-functional mix over the relevant historical period, and (2) consider what the place makes (dominant industries) and what it currently does (primary occupations).

Armed with this in-depth analysis, developers can craft a long-term strategy that identifies alternative avenues for movement up the learning curve to higher levels of personal income and skill. This strategy could, for example, suggest ways to change what local firms do, from routine production to precision production, as well as how to overcome impediments to further development, for instance, by resolving education or infrastructure problems.

Although Thompson's conceptualization was colored by his long tenure in large midwestern cities, his ideas are applicable to places of various size and location. The small metropolitan area currently engaged in routine production, for example, may be able to attract one large branch facility due to relatively low comparative costs and continue to recruit facilities to the location. These facilities should generate more tax revenues and strengthen the industrial real estate market. Subsequent industrial park development and infrastructure improvements should benefit existing local companies. Tighter labor markets may drive up wages and encourage the gradual upgrading of the labor force. Local companies may become more profitable and pay higher wages; new higher-wage businesses may blossom in the area. These companies could attract more firms providing supportive services and foster greater consumer demand for additional local goods and services. As a result, per capita incomes and income equality should increase.

Local economic developers who can think about development in these terms will be able to formulate *unique* local strategies reflecting what the local economy makes and does. With this orientation, developers can squarely face the major problems and opportunities of the area.

Columbus, Ohio

The important connections between wages and economic development explained by Thompson can be applied to Columbus, Ohio. Over the past five decades, different political-economic interests in Columbus have been trying to define the city's proper development path. Early battles raged between manufacturing and public administration interests. The latter interests prevailed and kept pushing strategies to promote service industries. Subsequently, wealthy locals got behind major projects, to give Columbus a more sophisticated image. Most projects have been located downtown or at Ohio State University. Others have been event-oriented, such as the Quincentenary

Columbus Day celebration in 1992. Critics argue that such image-building projects have come at the expense of improvement projects in poor and working-class neighborhoods. Recently, international opportunities have been considered as a way to compensate for stable employment in public administration.

At the root of these development options is the relative wage level for different occupational groups in the Columbus region. Local developers should analyze these levels comparatively, relative to those in the relevant competitive regions, rather than thinking in terms of a high-wage or low-wage strategy. For example, Columbus has successfully positioned itself as a lower-cost distribution center than other Midwestern cities. Spiegel followed its subsidiary, Eddie Bauer, from the Chicago area to Columbus to lower its labor and other costs. On the other hand, significant restructuring of manufacturers has been stimulated by competitive pressures in response to relatively high manufacturing wages. In general, the employment shift to service-sector jobs has helped sustain employment and output growth but has reduced the upward pressures on wages formerly stimulated by high-profit manufacturers. The power of labor unions and the wage roll-out effect have waned. Thus, sectoral shifts and productivity changes have influenced the occupational distribution and the resulting wage structure in Columbus.

Developers in the Southeast that currently compete with Columbus for high-wage manufacturers are changing their views on business climate. Previously, they associated low-wage labor, low taxes, and cheap land with a good business climate. Lower costs supported the absolute advantage of their area over others. Manufacturers were able to hire cheap labor in one location and sell their products to better-off workers in other locations.[9] Now, these developers advocate high-skill labor and best-practice technologies for local manufacturing. As a result, wages are increasing in these sectors.

Two directions are generally available in regions that are experiencing increases in prevailing wages. First, the labor force can become more skilled as the regional economy grows and, as a consequence, become more attractive to higher-skill, higher-wage industries. Second, labor can bargain with high-profit firms for a larger share of their profits. The former direction is more likely than the latter.

III. Conceptual Framework

The growth and development theories presented above can be synthesized and organized into a conceptual framework that can provide economic developers with a general frame of reference. This framework is not a growth or development theory. Rather, it organizes concepts using two "ideal types" of competitiveness that are related to the processes of growth and development. Either type of competitiveness can lead to long-term economic viability in one area compared to other areas.

Local economies grow, contract, or stagnate; they experience development or underdevelopment. Competitiveness is relevant to these processes at three different levels: the firm, the industrial sector, and the local economy as a whole. Firms become more competitive in several ways. According to trade theory, they expand their operations or intensify production through specialization and trade. Product-cycle theory emphasizes export growth that results in greater market share. Firms also restructure the organization to increase quality or reduce costs. These changes can spur both innovation and the formation of new companies.

Over time, industries experience cycles of growth and decline. Economic base theory, sector theory, and growth theory predict different ways by which local industrial sectors can continue to grow. Staple theory, product-cycle theory, entrepreneurship theories, and flexible-production theories suggest developmental activities that cause new industries to appear and others to disappear.

Local economies become more or less competitive as the result of (1) the aggregate impacts of local change at the firm or industry level, and (2) the resulting qualitative changes in economic structure that represent unique adaptations to external changes in technology, tastes, or institutional factors or adoption of proven ideas and practices. See Malizia (1986, 1990, 1996) for elaboration.

On the basis of the theories presented, then, competitiveness leads to growth and/or development. From the growth perspective, competitiveness may be considered efficient production. Economic base theory, neoclassical growth theory, interregional trade theory, and, to some extent, sector theory provide insights about the growth process that support efficient production. From the development perspective, competitiveness may be considered *flexible* production. Product-cycle theory, entrepreneurship, and flexible-production theories, as well as historically oriented theories such as staple theory, offer insights about economic development that help us understand how to achieve competitiveness through flexible production. Economic developers should view efficient production and flexible production as two strategic paths to follow in achieving greater competitiveness (see figure 11.1).[10]

At the firm level, efficiency increases with increasing productivity and the realization of internal economies of size or scale. At the industry level, efficient production is reflected in productivity gains, usually through an industrial complex or district that results in localization economies. For the entire local economy, efficient production depends upon competitive advantage

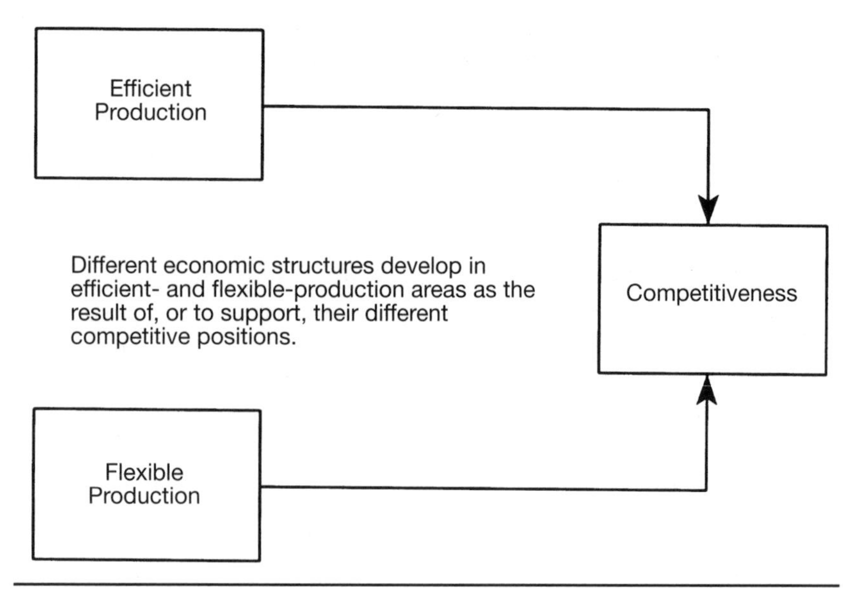

FIGURE 11.1
Two Paths to Competitiveness

which, at any given time, is a function of the functional specialization(s) of the area's industrial structure.[11] Efficient production can lead to sustained economic growth and, possibly, to greater stability.

Firms generating product or process innovations exemplify the flexible, adaptive producer. These firms achieve competitive strength through capable management and economies of scope. The latter are either internal to the firm or are formed through linkages to other firms. At the industry level, inter-firm networks may be formed to support flexible production. Local economies that are competitive due to flexible production realize urbanization economies which are generated by developmental business and professional services, efficient public services, and good intraurban transportation.[12]

It is important to understand that development in one part of the local economy, either in one firm or one industrial sector, need not have significant aggregate effects. Certainly, countervailing positive and negative changes can occur within one area, for two reasons. First, local firms make relatively independent decisions and control different types and amounts of resources. Second, external economic forces can move local industrial complexes and inter-firm networks in different directions. Because efficient or flexible production operates at the firm or industry level, a local economy can move toward competitiveness on both paths simultaneously. However, efficient firms and flexible firms influence their local environment differently, and the impacts tend to be cumulative and reinforcing. Eventually, the area's economic structure should evolve to become more supportive of one path or the other.

Thus, local economic structure can support either efficient production or flexible production. The fundamental attributes that enhance the area's capacity for long-term development have been presented in this book. They are summarized as follows:

Diversity: the variety of economic activity

Diverse metropolitan areas are more likely to attract innovative firms, both small and large. Competitive strength, which leads to long-term growth, is derived both from the relative stability economic diversity brings and from the flexibility inherent in local economies with multiple specializations. Narrowly specialized metropolitan areas can be just as competitive due to industrial complexes that contain highly productive firms. Usually, however, specialized areas are less stable.

Centrality: the economic importance of the place

The term *centrality* refers to the importance of economic activities and the related infrastructure found in the area. In view of the fact that strategic corporate planning directs production activity, corporate decision-making centers are more important than production centers. Places dense with corporate headquarters, money-center financial institutions, advanced business services, well-trained labor, and research universities are more important than places dominated by branch-plant production or single-establishment firms. Yet, subdominant areas are also competitive when their production and distribution channels remain efficient. As noted in Appendix 4.1, infrastructure usually complements dominant economic activities. Headquarters centers tend to have hub airports, good air travel connections, and advanced telecommunications, all of which enhance effective corporate control. Production centers tend to have good access to the interstate highway system, to port and railroad facilities, good air cargo facilities, and efficient municipal services.

Resilience: the effective response to adversity or opportunity

Skilled labor used in efficient production processes can generate price-competitive products for global markets. These efficient firms can adopt the

TABLE 11.1

Two Paths to Competitiveness:
Efficient Production and Flexible Production

	Efficient Production	*Flexible Production*
Firm Level:	Efficient Firms	Flexible Firms
Industry Level:	Industrial Complexes	Interfirm Networks
Labor Market Level:		
Diversity:	Narrowly Specialized	Multiple Specializations
Centrality:	Subdominant Areas	Dominant Areas
Resilience:	Adaptive Firms	Adaptive Innovative Firms
	Productive Industry Mix	Skilled Occupational Mix
Agglomeration Economies:	Localization	Urbanization

changes necessary to remain competitive by imitating successful practices or purchasing new technologies. Well-educated workers, talented managers, adaptable capital, and flexible technologies can produce specialized products that find niches in global markets. These flexible firms are adaptive and innovative. Thus, resilient areas have efficient or flexible firms that can "take a punch" or that can seize an opportunity which leads to higher levels of local economic development.

Finally, agglomeration economies can sustain development when they are more powerful than diseconomies, such as crime, congestion, declining environmental quality, and so on, which arise from growth.

In summary, the conceptual framework presented here is shown in table 11.1. The competitive differences among metropolitan areas reflect fundamental differences in the ability of individuals and firms to compete in the global economy. The competitive advantage of areas relying on efficient production is based on the functional specializations and localization economies built on and related to firm-level economies of scale or size. The competitive advantage of areas manifesting flexible production depends on the flexibility, capacity for adaptation, responsiveness, and creativity of their firms. These areas maintain one or more evolving specializations which are supported primarily by urbanization economies. Different economic structures are reflected in different levels of diversity, centrality, and resilience. Efficient- and flexible-production areas have different economic structures as the result of, and for the support of, their different competitive positions.

Practical Applications

The conceptual framework should help local economic developers understand more fully the labor market area in which they work. They should examine each element of the framework to describe the economic base accurately. Furthermore, developers can study the jurisdictions *within* the regional economy to find structural differences that may result in different competitive advantages. The regional approach makes sense when wage levels, skills and occupations, developmental services, tax structure, and so forth vary within the labor market area. Developers can pursue growth strategies and developmental strategies at the same time but for different localities in the region. For example, the core area may be more attractive for service industries or flexible firms, while the peripheral areas may be more attractive to manufacturers or firms seeking relatively low-cost locations.

More fundamentally, developers can use the framework to identify *comparable metropolitan regions* or labor market areas that compete with their locality. For example, the Raleigh–Durham metropolitan economy is anchored by three research universities, Research Triangle Park, and state government. The economy is research- and development-oriented and, to a lesser extent, administration-oriented. It has multiple, functional specializations in government services and administration, educational services, health services, business services, and electronics manufacturing. The most comparable metropolitan economy appears to be Austin, Texas. Other possible "comparables" include the following areas: Columbus, Ohio; Nashville, Tennessee; Norfolk–Virginia Beach, Virginia; Richmond, Virginia; Salt Lake City, Utah; and Tucson, Arizona. Raleigh–Durham competes with these areas to a much greater extent than with Charlotte or Greensboro, North Carolina. Economic developers may find it worthwhile to monitor economic trends in their comparable metropolitan areas and to pay attention to economic development practice in these areas.[13]

The two strategic paths support the *business development activities* of economic developers. Business attraction, expansion, or creation activities are legitimate ways to facilitate job creation and investment.[14] Economic developers traditionally have embraced the efficient production path to formulate their attraction and expansion programs. They work to reduce the capital and operating costs for recruited or expanding companies. They promote workforce preparedness, adequate basic infrastructure, and favorable business climate to cost-conscious firms.

Economic developers supporting flexible production attend primarily to new firms or to expanding local firms. The quality of education and other public services, maintenance and repair services, and Thompson's developmental services is important to flexible producers. Vocational training that emphasizes general analytical skills and computer literacy is more essential than narrowly defined, "customized" training programs. Opportunities for course work or joint research at local colleges and universities are valued by flexible firms.

These activities focus on ways to increase the competitiveness of strategically important firms. These business development activities offer greater promise than do naive export promotion or import substitution, either of which may inadvertently reduce local competitiveness.[15]

The relative attractiveness of most local economies is determined primarily by the competitiveness of its industries and firms because households are assumed to follow firms (employment opportunities) more than they are to do the reverse. However, the conceptual framework can be extended to offer insights about the *location of households,* as well as firms. In fact, relative attractiveness is the general feature that determines the long-term performance of metropolitan economies. Individuals who are free to live where they choose, either temporarily (tourists) or permanently (retirees and wealthy individuals), gauge the quality of life offered by different places. Thus the relative attractiveness of places to footloose households is determined not by competitiveness but by quality of life.[16]

For this application, quality of life may be defined as including housing, education, health care, natural amenities, cultural amenities, physical safety and security, and accessibility. In evaluating quality of life, households assess the range of private and public goods and services available in each area. This range is a function of where the area is located in the urban hierarchy. In essence, quality of life depends on the quality of these goods and services, as well as their cost. Firms seeking locations that enhance their competitiveness are concerned with the same general factors—the availability, quality, and cost of requisite goods and services found in different places.

Some places are attractive because they are relatively cheap, enabling the footloose household with limited resources or fixed incomes to enjoy a reasonable lifestyle. Other places are attractive because they offer relatively high-quality living, but at premium prices. For example, along these lines, resort communities compete for tourists, retirees, wealthy individuals, and so forth. Some offer decent services at bargain prices, while others are expensive but deliver services of high quality. As with locales whose relative attractiveness is determined by competitiveness (efficient production *or* flexible production), one resort community cannot easily "wear two hats." In general, the performance of a metropolitan area depends on its relative attractiveness—either quality of life for households or competitiveness for firms. Some metropolitan areas will attract cost-conscious households and efficiency-oriented firms; others will attract households more concerned with quality and firms more interested in achieving flexibility.

Entrepreneurs tend to be footloose. They can found ventures in many different locations. All they need is physical and electronic access from those locations to their regional, national, or international markets. Entrepreneurs select locations by considering both lifestyle and economic factors. They try to optimize (1) quality of life for themselves and their family and (2) competitiveness for their business.

In conclusion, economic developers are in a competitive business. They often compete by offering the same programs as the competition or by trying to outbid the competition with incentives. Neither approach requires sophisticated understanding of the local economy. The theories presented as synthesized in the efficient and flexible production paths to competitiveness can elevate the thinking about development strategies. Economic developers are working in unique local economies that compete with a specific set of comparable areas; therefore, they need unique strategies based on original analysis and creative problem-solving.[17]

IV. Conclusions

Chapters 1 through 10 demonstrated insights about economic development and economic growth deducible from a synthetic treatment of theories. Most economic developers will want to construct an eclectic theory of development by selectively drawing on these ideas. Yet, while a carefully formulated synthesis of theories can serve an important purpose, the developer can realize real benefits from applying each theory separately. By exploring each theoretical perspective carefully and fully, while at the same time being mindful of the specific theoretical questions and empirical problems the perspective is attempting to address, the developer is more likely to discover particular insights that clarify the process of local economic development. If different theories support the same development strategies, the priority of these strategies may be elevated as a result. As is argued in Part I, *without theory, the economic developer pursues politically expedient strategies with professionally accepted techniques but has no way to build a defensible, independent basis for understanding and action. With theory, the developer can first understand the threats and opportunities facing the local economy and can fashion unique strategies that address its strengths and weaknesses*. Instead of politics dictating only partly relevant development efforts, economically significant strategies may be adapted to meet political realities. In this sense, a good understanding of theory will enhance the economic developer's creativity and ability to design more effective solutions to economic problems.

As a practical matter, economic developers employed by a particular jurisdiction within a given metropolitan or nonmetropolitan area should try to understand their local economy in two stages. In the first stage, the developer should use theory to inform the analysis of the appropriate spatial economic unit within which the jurisdiction is a part. As the applications make clear, the subject of this book—understanding local economic development—pertains to labor market/metropolitan areas. Since labor markets function as local economies, it makes sense to analyze them as whole units. Although some local economic developers do, indeed, focus on the entire local economy (being employed by local Chambers or area development commissions set up for this purpose), most are hired by political entities that constitute only a part of the labor market/metropolitan area, thus leading them to consider only that subset of the larger region that is their primary jurisdiction. This approach must be avoided, since regional development theories have little relevance when divorced from the context of functional economic regions. After adopting a broader, region-wide perspective, the developer should identify the most relevant economic development strategies that address the area's uniqueness. Then, in the second stage of analysis, the developer can proceed to "step-down" the strategies by finding particular ones or adaptions of others that best match the features of the subject jurisdiction.

As noted in the previous section, differentiation among strategies within labor market areas is easiest to visualize when the labor market contains one large central city and a periphery of smaller communities and rural areas. Yet, strategies may be differentiated in more complex regions, which contain several large jurisdictions or numerous smaller ones. Each jurisdiction can distinguish its specific economic development strategy on the basis of within-region differences in factor costs, accessibility, amenity levels, business climate, economic base, and structural development attributes. Some jurisdictions in the region may specialize in residential development; some may be "antigrowth" and thus prefer not to support economic development.

In reality, considerable competition often exists among local economic developers hired by different communities that are part of the same labor market area. This competition frequently drives development professionals working in the same area to pursue the same strategies and to target the same existing industries or prospects, which, more often than not, reduces everyone's effectiveness. Just as the theories presented in this book can reduce *inter*regional competition among economic developers by helping them understand the unique features of their labor market area, the theories can also help reduce *intra*regional competition. First, economic developers from the same labor market area can form discussion forums and, together, try to arrive at a common understanding of their local economy. Second, these developers can analyze the variations in the economic landscape of the local jurisdictions, with a view to understanding their interrelationships and interdependencies. This, in turn, will allow them to particularize strategies to their jurisdiction that make sense at the labor market level. For example, the largest jurisdiction or a strategically located smaller jurisdiction may offer the labor skill mix and amenity package attractive to specialized service firms, while other jurisdictions may be able to provide the industrial space and skilled labor required by cost-conscious manufacturing firms. As long as job creation remains the central concern, such region-wide cooperation should be attractive, since employment opportunities within the labor market area, by definition, are available to all.

Even with this approach, interjurisdictional competition for economic development may remain. Unfortunately, tax base expansion is often emphasized; in these instances, developers are hard-pressed not to compete for the same ratables.[18] Perhaps this would not be the case if elected officials, developers, and citizens had a more complete understanding of the dynamics of a regional economy and the microbehavioral influences on the firms and households that make it up. In reality, the industrial location and expansion process will respond to the differences within labor markets/metropolitan areas even if local economic developers are unaware of them or if they choose to compete directly with their neighbors. Expansion and location decisions are very important to both large and small firms, and these firms are usually able to find the sites and jurisdictions that best meet their needs.

Most books and articles about economic development in the United States pose, either explicitly or implicitly, one preferred way to view development or one set of favored strategies. Economic developers have been admonished to cease recruiting industry and instead to create manufacturing networks, build incubator facilities, capitalize venture funds, promote tourism, attend to small business or existing industries, cultivate entrepreneurship, and encourage international exports—to mention some of the more popular suggestions. This advocacy of particular viewpoints on strategies is contrary to the message of this book. For local economic developers, the real challenge is to arrive at strategies that are uniquely suited to their local economy and jurisdiction. These strategies should be most similar to those of comparable labor market areas, that is, to areas with similar characteristics and economic locations in the global system. Still, even in these instances, specific strategic differences will emerge.

With a genuine understanding of the local economic development process, economic developers can promote their region's competitiveness by facilitating more work, better work, or new work. They may be able to subordinate near-term growth and build the necessary physical, business, and political infrastructure to pursue the long-term objective of economic development. They can work to improve the overall milieu of the area or target efforts to particular firms, workers, industries, or subareas. The key to success lies in determining how and from where specific strategies and programs are derived. In places where strategies rationalize existing programs, techniques, and political expediencies, success will depend more on luck than on skill or effort. In towns, cities, and regions where strategies flow from deep insights into the structure and function of the local economy, economic development efforts stand a far greater chance of success.

Notes

1. Although the industry, occupation, enterprise, and product dimensions of regional economies are all relevant, these theories tend to introduce or emphasize just one or two of these categories. Growth theories emphasize industry mix, especially the leading export industries. Product-cycle theory focuses on the product dimension, while entrepreneurship theories introduce the enterprise or company dimension. In general, growth theories tend to be more macrolevel and deterministic, whereas development theories are more microlevel and mediated (see chapter 2). Theories of innovation focus on a locality's ecological structure—its diversity and centrality, as well as its business climate and quality of life. These concepts are introduced in chapter 2 and addressed more fully in the conceptual framework in chapter 11.

2. Thompson sees all aspects of development in comparative terms. Indicators of development are always comparative, either across regions or over time in one place.

3. The concepts of breadth and depth appear to be consistent with Shapero's (1981) argument about the capacity of local economies to create new work in response to external change. *Breadth* is shorthand for diversity of economic base. *Depth* appears to support the economy's ability to change or, in Shapero's terms, to demonstrate resilience.

The only deficiency of the statement is the assumption that city size functions as a causal factor. More likely, size is correlated with, or the outcome of, more basic growth and development factors: economic diversity, centrality, and resilience. See the discussion of city size in Appendix 4.1.

4. Thompson seems to exaggerate the advantages of city size and to underestimate the importance of large corporate organizations that tend to be headquartered in large cities. Other theorists recognize the major influence of large multilocational firms, for example, Pred (1976), Markusen (1985), and Malecki (1997).

5. Rosabeth Moss Kanter (1995) presents a compatible, but less well structured, approach to attaining competitiveness in the global economy. Cities she describes as "thinkers" are essentially R&D-oriented places. "Makers" are places that focus on precision production, either in manufacturing or service industries. Other cities are "traders" that focus on international finance and communication. For trade-oriented cities, Kantor has combined the historical competitive advantage of port cities—as physical and commercial breaks in transportation—with their current role in the urban hierarchy as headquarters cities. Although advantaged by trade-oriented infrastructure, these places currently function as administrative/command and control centers.

6. Although this scheme is conceptually attractive, the empirical measures of occupational specialization are rather weak.

7. Thompson does not identify developmental services precisely. The following SIC industry groups are examples of largely developmental service activities: 451, Air Transportation, Scheduled, and Air Courier Services; 458, Airports, Flying Fields, and Airport Terminal Services; 616, Mortgage Bankers and Brokers; 621, Security Brokers, Dealers, and Flotation Companies; 635, Surety Insurance; 679, Miscellaneous Investing; 731, Advertising; 733, Mailing, Reproduction, Commercial Art and Photography, and Stenographic Services; 735, Miscellaneous Equipment Rental and Leasing; 736, Personnel Supply Services; 737, Computer Programming, Data Processing, and Other Computer Related Services; 738, Miscellaneous Business Services; 811, Legal Services; 822, Colleges, Universities, Professional Schools, and Junior Colleges; 871, Engineering, Architectural, and Surveying Services; 872, Accounting, Auditing, and Bookkeeping Services; 873, Research, Development, and Testing Services; 874, Management and Public Relations Services. Consider also 354, Metalworking Machinery and Equipment; and 361, Electric Transmission and Distribution Equipment in the manufacturing sector.

8. Public officials are supposed to have longer time horizons than corporate officers; yet, often, the reverse is true. The social discount rate—which, in theory, is assumed to be lower than the private discount rate—generally is higher in practice.

9. In the long term, however, there is a contradiction between low-wage employees and high-income customers. Just as all regions cannot increase exports and decrease imports simultaneously, all firms cannot buy labor cheaply from households and also sell expensive products to these households. Gradually, higher wages lead to better levels of living in a place, and this represents economic progress. Thus, economic developers should consider higher wages a goal that is worth pursuing.

10. Economic developers may find the following contrast useful in understanding the distinctions between efficient production and flexible production. Efficient production arose from the mass-production approach to economic growth. With this mode of production, factor inputs were increasingly refined and specialized: specially trained labor, narrowly defined production tasks, highly specialized machinery and equipment, specifically organized production lines. As a result, uniform products were produced in high volume at increasingly lower unit costs and were sold in mass consumer markets.

Flexible production emphasizes continuous improvement of quality. Product and process innovation are an integral part of the production process. Factor inputs are increasingly generalized. In other words, workers have the ability to perform many different mental and manual tasks; machinery and equipment can be used in a variety of ways. Computer-aided design and manufacturing replace the "tooling" formerly used in modifying machines for the mass production of products. With these "general" factor inputs, production runs can be short, or they can represent batch processing, yet still be cost-effective. As a result, specialized products can be targeted to niche markets, or standard products can be customized to meet specific customer needs.

One should remember that these distinctions are ideal types rather than explanations of reality. Actual modern production processes are more complex and often combine efficient and flexible production. For example, mass production increasingly has become customized with computer-linked, point-of-sale, and just-in-time inventory systems. Automobiles, whose production created the popular image of mass production with Ford's Model T, are now often purchased before they are produced, to meet the customer's specifications. On the other hand, niche products are uniform within the targeted market segment and remain sensitive to unit production costs.

11. For many decades, geographers, economists, sociologists, and regional scientists have been concerned with these fundamental attributes in the context of the urban hierarchy. Functional specialization has been researched extensively as an important determinant of local economic performance. In the classic *Regions, Resources and Economic Growth,* Perloff et al. (1960) developed shift–share analysis to analyze the relationship between industry mix and employment growth. They viewed functional specialization as the driving force in the metropolitan economy.

12. Agglomeration economies also help us understand flexible production when responsiveness, resilience, creative problem-solving, or innovation of local industries impact the overall local economy. See Appendix 4.1.

13. Metropolitan areas often use the U.S. average as the norm—for example, in measuring location quotients. Cities in the same state or region are often compared. These comparisons are based on the *geographic* location of the area, which reflects proximity, political boundaries, and economic history. The conceptual framework suggests that geographic location is inferior to *economic* location. Metropolitan areas with similar economic locations compete because they play similar roles in the global economy. For example, the larger metropolitan areas in the states of Ohio or South Carolina may have similar history, culture, climate, or natural resources due to proximate geographic location. As indicated in table 4.1, their economic roles vary: Greenville is a manufacturing center, Charleston is industrial-military (although more tourist- and technology-oriented in the 1990s), and Columbus, Ohio, is a government-educational center. Cleveland, Cincinnati, and Columbus play different roles in the global system.

Economic developers can use available economic statistics to track the performance of their local economy and its comparable metropolitan areas. It would be useful to compile and compare the following measures for these areas from 1970 to the present: per capita personal income, median household income, unemployment rate, percent of households in poverty (or a measure of income inequality), income and/or employment instability. The latter should be measured as the variation from the long-term income or employment trend. See Kort (1981) for further discussion of this instability measure.

14. However, retention activities are not supported because they usually involve subsidizing uncompetitive firms that have little chance of long-term survival. Developers should consider restructuring instead of retention. Restructuring may involve introducing new process technologies, retraining workers, reducing levels of management, changing suppliers, out-sourcing internal operations, revising marketing strategies, refinancing operations, and so on. The result of restructuring should be more competitive local (efficient or flexible) firms, although the total employment at these firms may decline.

15. These two paths to competitiveness also can help developers understand the apparently contradictory surveys designed to assess state business climate. The traditional surveys conducted during the 1970s and 1980s rated state business climate from the efficient production perspective. The availability and cost of local inputs received priority. The Corporation for Enterprise Development's score card, on the other hand, rates states according to the quality of inputs, reflecting the perspective of flexible production. The criticism that the former surveys measure "the cost of everything but the value of nothing" misses the essential point. To many firms, industries, and places, the relative cost of necessary inputs is more important than considerations of quality.

16. A growing percentage of households in the United States are footloose. The tourist industry, which is an important part of many local economies, serves people who are temporarily footloose. Similarly, students, military personnel, and even prison inmates constitute transient yet important market segments in many places. Increasingly, the self-employed or corporate employees use computers and telecommunications to enable them to live where they choose. As a result, many localities, particularly smaller communities and nonmetropolitan areas, may be able to grow and develop by attracting footloose households.

The importance of footloose households will increase as the U.S. population ages. Retired persons are becoming a growing economic force. Their substantial purchasing power will provide the economic base for some metropolitan areas.

17. There are no specific discussion questions at the end of this chapter. Instead, readers are expected to use the synthesis, conceptual framework, and applications to formulate unique development strategies on the basis of a better understanding of their local economy.

To facilitate this effort, we recommend the risk–reward framework used in portfolio theory. Each development strategy should be assigned a risk score and a reward score in terms of how well each achieves local economic development objectives. Risk scores can reflect a rank ordering of strategies or, more simply, they can be assigned as high-risk, medium-risk, or low-risk. Reward scores can also be the rank ordering of strategies or reflect the expected outcomes measured of each strategy in terms of jobs created, dollars invested, tax revenues generated, etc. All strategies can then be mapped in two dimensions, with degree of risk and reward as the axes. Strategies that achieve the highest reward at the lowest risk levels comprise the optimum portfolio of development strategies. For further discussion and illustration, see Malizia (1985, pp. 44–49).

18. The advantages of regional tax base sharing are obvious in this regard, but the practice is not yet widespread, nor is it legally permitted in most states.

Bibliography

Works Cited

Ács, Z. 1996. Small firms and economic growth. In P. Admiraal, ed., *Small firms in the modern economy*. Cambridge: Blackwell.

Ács, Z., and Audretsch, D. 1987. Innovation, market structure and firm size. *Review of Economics and Statistics* 69: 657-675.

AEDC (American Economic Development Council). 1984. *Economic development today: a report to the profession*. Schiller Park, Ill.: AEDC.

AEDC (American Economic Development Council). 1991. R. Swager, ed. *Economic development tomorrow: a report from the profession*. Rosemont, Ill.: AEDC.

Alchain, A. 1950. Uncertainty, evolution, and economic theory. *Journal of Political Economy* 58: 211-221.

Allen, D., and McClusky, R. 1990. Structure, policy, services, and performance in the business incubator industry. *Entrepreneurship theory and practice* 1990: 61-77.

Alonso, W. 1964. Location theory. In J. Friedmann and W. Alonso, eds., *Regional development and planning*. Cambridge: MIT Press.

Alonso, W. 1990. From Alfred Weber to Max: the shifting of regional policy. In M. Chatterji and R. Kuenne, eds., *Dynamics and conflict in regional structural change*. New York: New York University Press.

Amin, A., and Robins, K. 1990. The re-emergence of regional economies? The mythical geography of flexible accumulation. *Environment and Planning D: Society and Space* 8: 7-34.

———. 1991. These are not Marshallian times. In R. Camagni, ed., *Innovation networks: spatial perspectives*. London: Belhaven.

Amos, O. 1990. Growth pole cycles: a synthesis of growth pole and long wave theories. *Review of Regional Studies* 20: 37-48.

Anderson, G. 1994. Industry clustering for economic development. *Economic Development Review* 12: 26-32.

Antonelli, C. 1990. Induced adoption and externalities in the regional diffusion of information technology. *Regional Studies* 24: 31-40.

Armstrong, H., and Taylor, J. 1978. *Regional economic policy and its analysis*. Oxford: Philip Allan.

———. 1985. *Regional economics and policy*. Deddington, Oxford: P. Allan.

Arndt, H. 1981. Economic development: a semantic history. *Economic Development and Cultural Change* 29: 457-466.

Asheim, B. T. 1996. Industrial districts as "learning regions": a condition for prosperity. *European Planning Studies* 4, 4, 379-400.

Ayau, M. 1983. Basics of comparative advantage aren't so hard to learn. *Wall Street Journal*, October 20.

Bal, F., and Nijkamp, P. 1998. Exogenous and endogenous spatial growth models. *Annals of Regional Science* 32, 1: 63-89.

Barnet, R., and Mueller, R. 1974. *Global reach: the power of the multi-national corporation*. New York: Simon and Schuster.

Bauer, P., and Yamey, B. 1951. Economic progress and occupational distribution. *Economic Journal* 61: 741-755.

Baumol, W. 1968. Entrepreneurship in economic theory. *American Economic Review* 58: 64-71.

———. 1983. Toward operational models of entrepreneurship. In J. Ronen, ed., *Entrepreneurship*. Lexington, Mass.: D.C. Heath, 29-48.

Beale, C. 1979. A further look at nonmetropolitan population growth since 1970. In *The demographic shift toward rural areas*. Washington, D.C.: CBO. Originally published in *American Journal of Agricultural Economics* 58 (1976).

Beckman, M. 1958. City hierarchies and the distribution of city size. *Economic Development and Cultural Change* 6: 243-248.

Beguin, H. 1985. A property of the rank-size distribution and its use in an urban hierarchy context. *Journal of Regional Science* 25: 437-441.

Bellini, N. 1996. Italian industrial districts: evolution and change. *European Planning Studies* 4, 1: 3-4.

Benvignati, A. 1982. Interfirm adoption of capital-goods innovations. *Review of Economics and Statistics* 64: 330-335.

Bergman, E., and Feser, E. 1998. Lean production systems in regions: conceptual and measurement requirements. Forthcoming, *Annals of Regional Science*.

Bergman, E.; Feser, E.; and Scharer, J. 1995. *Modern practices and needs: North Carolina's transportation equipment manufacturers*. Chapel Hill, N.C.: U.N.C. Institute for Economic Development.

Bergman, E.; Feser, E.; and Sweeney, S. 1996. *Targeting North Carolina manufacturing: understanding the state's economy through industrial cluster analysis*. Chapel Hill, N.C.: U.N.C. Institute for Economic Development.

Berry, B. 1972. Hierarchical diffusion: the basis of developmental filtering and spread in a system of growth centers. In N. Hansen, ed., *Growth centers in regional economic development*. New York: Free Press, 108–138.

Berry, B.; Conkling, E.; and Ray, D. 1997. *The global economy in transition*. 2nd ed. Upper Saddle River, N.J.: Prentice Hall.

Best, M. 1990. *The new competition: institutions of industrial restructuring*. Cambridge: Polity Press.

Beyers, W., and Alvine, M. 1985. Export services in postindustrial society. *Papers in Regional Science* 57: 33–45.

Beyers, W., and Lindahl, D. 1996. Explaining the demand for producer services. *Papers in Regional Science* 75: 351–374.

Billings, R. 1969. The mathematical identity of the multipliers derived from the economic base model and the input-output model. *Journal of Regional Science* 9: 471–473.

Bingham, R., and Mier, R., eds. 1993. *Theories of local economic development*. Newbury Park, Calif.: Sage Publications.

———. 1997. *Dilemmas of urban economic development: issues in theory and practice*. Thousand Oaks, Calif.: Sage Publications.

Binks, M., and Vale, P. 1990. *Entrepreneurship and economic change*. New York: McGraw-Hill.

Birch, D. 1979. *The job generation process*. Cambridge, Mass.: MIT Program on Neighborhood and Regional Change.

———. 1981a. Generating new jobs: are government incentives effective? In R. Friedman and W. Schweke, eds., *Expanding the opportunity to produce: revitalizing the American economy through new enterprise development*. Washington, D.C.: Corporation for Enterprise Development, 10–16.

———. 1981b. Who creates jobs? *The Public Interest* 65: 3–14.

———. 1987. *Job creation in America*. New York: Free Press.

Blair, J. 1991. *Urban and regional economics*. Homewood, Ill.: Richard Irwin.

Blakely, E. 1994. *Planning local economic development: theory and practice*. Thousand Oaks, Calif.: Sage Publications.

Blaug, M. 1978 [1968]. *Economic theory in retrospect*. 3rd ed. Homewood, Ill.: Richard D. Irwin.

———. 1997. *Economic theory in retrospect*. New York: Cambridge University Press.

Blumenfeld, H. 1955. The economic base of the metropolis, *Journal of the American Institute of Planners* 21: 114–132.

Boekholt, P. 1997. The public sector at arm's length or in charge? Towards a typology of cluster policies. Paper presented at the OECD Workshop on Cluster Analysis and Cluster Policies, Amsterdam, Netherlands, October 9–10, 1997.

Borts, G., and Stein, J. 1964. *Economic growth in a free market*. New York: Columbia University Press.

Bosworth, B., and Rosenfeld, S. 1993. *Significant others: exploring the potential of manufacturing networks*. Chapel Hill, N.C.: Regional Technology Strategies.

Bottomore, T. 1966. *Classes in modern society*. New York: Pantheon Books.

Boulding, K. 1956. *The image: knowledge in life and society*. Ann Arbor: University of Michigan Press.

Branson, W. 1989. *Macroeconomic theory and policy*. 3rd ed. New York: Harper and Row.

Bronfenbrenner, M. 1979. *Macroeconomic alternatives*. Arlington Heights, Ill.: AHM Publishing.

Bruno, A., and Tyebjee, T. 1982. The environment for entrepreneurship. In C. Kent, D. Sexton, and K. Vesper, eds., *Encyclopedia of entrepreneurship*. Englewood Cliffs, N.J.: Prentice-Hall, 288–307.

Casetti, E.; King, L.; and Odland, J. 1971. The formalization and testing of concepts of growth poles in a spatial context. *Environment and Planning* 3: 377–382.

Casson, M. 1987. Entrepreneur. In J. Eatwell, M. Milgate, and P. Newman, eds., *The new Palgrave dictionary of economics*, vol. 2. London: Macmillan, 151–153.

Chichilnisky, G. 1997. What is sustainable development? *Land Economics* 73: 467–491.

Chisholm, M. 1990. *Regions in recession and resurgence*. London: Unwin Hyman.

Chong-Yah, L. 1991. *Development and underdevelopment*. London: Longman Group.

Clark, C. 1960. *The conditions of economic progress*. New York: Macmillan.

Clinton, W. 1997. *Economic report of the President*. Washington, D.C.: Government Printing Office.

Coase, R. 1937. The nature of the firm. *Economica* 4: 386–405.

Cobb, J. 1993. *The selling of the South: the southern crusade for industrial development, 1936–1980*. Baton Rouge: Louisiana State University Press.

Conroy, M. 1975. *Regional economic diversification*. New York: Praeger.

Cooper, H. 1994. Southern comfort: Tupelo, Miss., concocts an effective recipe for economic health. *Wall Street Journal*. March 3: a1.

Courchene, T., and Melvin, J. 1987. A neoclassical approach to regional economics. In B. Higgins and D. Savoie, eds., *Regional economic development*. Boston: Unwin Hyman.

Cox, K. 1988. Locality and community in the politics of local economic development. *Annals of the American Association of Geographers* 78: 307-325.

Czamanski, S. 1973. *Regional and interregional social accounting*. Lexington, Mass.: Lexington Books.

———. 1974. *Study of clustering of industries*. Halifax, Nova Scotia: Institute of Public Affairs.

———. 1976. *Study of spatial industrial complexes*. Halifax, Nova Scotia: Institute of Public Affairs.

Czamanski, S., and Ablas, L. 1979. Identification of industrial clusters and complexes: a comparison of methods and findings. *Urban Studies* 16: 61-80.

Daly, H. 1991. *Steady state economics*. Washington: Island Press.

Darwent, D. 1969. Growth poles and growth centres in regional planning: a review. *Environment and Planning* 1: 5-31.

Demery, D., and Demery, L. 1970. Statistical evidence of balanced and unbalanced growth: comment. *Review of Economic Statistics* 52: 108-110.

———. 1973. Cross-section evidence for balanced and unbalanced growth. *Review of Economic Statistics* 55: 459-464.

Doeringer, P., and Terkla, D. 1995. Business strategy and cross-industry clusters. *Economic Development Quarterly* 9: 225-237.

Domar, E. 1946. Capital expansion, rate of growth, and employment. *Econometrica* 14: 137-147.

———. 1947. Expansion and employment. *American Economic Review* 37: 34-55.

Dome, T. 1994. *History of economic theory: a critical introduction*. Hants, England: Edward Elgar.

Dorfman, R. 1991. Review article: economic development from the beginning to Rostow. *Journal of Economic Literature* 29: 573-591.

Dubini, P. 1988. The influence of motivations and environment on business start-ups: some hints for public policies. *Journal of Business Venturing* 4: 11-26.

Dunne, T. 1994. Plant age and technology use in U.S. manufacturing industries. *Rand Journal of Economics* 25: 488-499.

Elliot, J. 1980. Marx and Schumpeter on capitalism's creative destruction. *Quarterly Journal of Economics* 95: 45-68.

———. 1985. Schumpeter's theory of economic development and social change: exposition and assessment. *International Journal of Social Economics* 12: 6-33.

Erickson, R. 1992. Trade, economic growth, and state export promotion programs. In R. McGowan and E. Ottensmeyer, eds., *Economic development strategies for state and local governments*. Chicago: Nelson-Hall.

Erickson, R., and Leinbach, T. 1979. Characteristics of branch plants attracted to nonmetropolitan areas. In R. Lousdale and H. Seyler, eds., *Nonmetropolitan industrialization*, 57-78. New York: Halsted Press; Washington: V. H. Winston.

Ettlinger, N. 1994. The localization of development in comparative perspective. *Economic Geography* 70, 2: 144-166.

———. 1997. An assessment of the small-firm debate in the United States. *Environment and Planning A* 29: 419-442.

Farness, D. 1989. Detecting the economic base: new challenges. *International Regional Science Review* 12: 319-328.

Feser, E. 1998a. Enterprises, external economies, and economic development. *Journal of Planning Literature* 12, 3: 283-302.

———. 1998b. Old and new theories of industry clusters. In M. Steiner, ed., *Clusters and regional specialisation*, 18-40. London: Pion.

Feser, E., and Bergman, E. 1998. National industry clusters: frameworks for state and regional development policy. Forthcoming, *Regional Studies*.

Feser, E., and Sweeney, S. 1998. Industrial complexes revisited: a test for coincident economic and spatial clustering. Working paper, Center for Urban and Regional Studies, University of North Carolina at Chapel Hill.

Fisher, A. 1933. Capital and the growth of knowledge. *Economic Journal* 43: 379-389.

Flammang, R. 1979. Economic growth and economic development: counterparts or competitors. *Economic Development and Cultural Change* 28: 47-62.

———. 1990. Development and growth revisited. *Review of Regional Studies* 20: 49-55.

Foss, N., and Knudsen, C. 1995. *Towards a competence theory of the firm*. London and New York: Routledge.

Frey, D. 1989. A structural approach to the economic base multiplier. *Land Economics* 65: 352-358.

Friedmann, J. 1966. *Regional development policy: a case study of Venezuela*. Cambridge, Mass.: MIT Press.

———. 1972. A generalized theory of polarized development. In N. Hansen, ed., *Growth centers in regional economic development*. New York: Free Press, 82-107.

———. 1986. The world city hypothesis. *Development and Change* 17: 69-84.

———. 1995. Where we stand: a decade of world city research. In P. L. Knox and P. J. Taylor, eds., *World cities in a world system*. Cambridge: Cambridge University Press, 21-47.

Friedmann, J., and Alonso, W. 1964. *Regional development and planning, a reader*. Cambridge, Mass.: MIT Press.

Friedmann, J., and Weaver, C. 1979. *Territory and function: the evolution of regional planning*. Berkeley: University of California Press.

Gaile, G. 1980. The spread-backwash concept. *Regional Studies* 14: 15-25.

Galbraith, J. 1987. *Economics in perspective*. Boston: Houghton Mifflin.

Garnick, D. 1970. Differential regional multiplier models. *Journal of Regional Science* 10: 35-47.

Gertler, M. 1988. The limits to flexibility: comments on the post-Fordist vision of production and its geography. *Transactions of the Institute of British Geographers* 13: 419-432.

———. 1993. Implementing advanced manufacturing technologies in mature industrial regions: towards a social model of technology production. *Regional Studies* 27: 665-680.

Ghali, M.; Akihama, M.; and Fujiwara, J. 1978. Factor mobility and regional growth. *Review of Economics and Statistics* 60: 78-84.

———. 1981. Models of regional growth: an empirical evaluation. *Regional Science and Urban Economics* 11: 175-190.

Gianaris, N. 1978. *Economic development: thought and problems*. North Quincy, Mass.: Christopher Publishing House.

Giarratani, F., and Soeroso, F. 1985. A neoclassical model of regional growth in Indonesia. *Journal of Regional Science* 25: 373-382.

Goldstein, H., and Luger, M. 1990. Science/technology parks and regional development theory. *Economic Development Quarterly* 4: 64-78.

———. 1991. *Technology in the garden*. Chapel Hill: University of North Carolina Press.

Gore, C. 1984. *Regions in question: space, development theory, and regional policy*. New York: Methuen.

Granovetter, M. 1985. Economic action and social structure: the problem of embeddedness. *American Journal of Sociology* 91: 481-510.

Gray, M.; Golob, E.; and Markusen, A. 1996. Big firms, long arms, wide shoulders: the "hub-and-spoke" industrial district in the Seattle region. *Regional Studies* 30: 651-666.

Greenwood, M. 1975. Research on internal migration in the USA: a survey. *Journal of Economic Literature* 13: 397-433.

Greytak, D. 1975. Regional consumption patterns and the Heckscher-Ohlin trade theorem. *Journal of Regional Science* 15: 39-45.

Grossman, G. 1996. Introduction. In G. Grossman, ed., *Economic growth: theory and evidence*. Cheltenham, U.K.: E. Elgar Publications.

Hale, C. 1967. The mechanics of the spread effect in regional development. *Land Economics* 43: 433–444.

Hamberg, D. 1971. *Models of economic growth*. New York: Harper and Row.

Hansen, N. 1967. Development pole theory in a regional context. *Kyklos* 20: 709–725.

———. 1971. *Intermediate-size cities as growth centers*. New York: Praeger.

Harrington, J. W., and Warf, B. 1995. *Industrial location: principles, practice, and policy*. London: Routlege.

Harrison, B. 1992. Industrial districts: old wine in new bottles? *Regional Studies* 26, 5: 469–483.

———. 1994. *Lean and mean: the changing landscape of corporate power in the age of flexibility*. New York: Basic Books.

Harrison, B.; Kelley, M.; and Gant, J. 1996a. Specialization vs. diversity in local economies: the implications for innovative private sector behavior. *Cityscape* 2: 61–93.

———. 1996b. Innovative firm behavior and local milieu: exploring the intersection of agglomeration, firm effects, and technological change. *Economic Geography* 72: 233–258.

Harrod, R. 1939. An essay in dynamic theory. *Economic Journal* 49: 14–33.

Hawley, A. 1950. *Human ecology: a theory of community structure*. New York: Ronald Press.

Hayward, D. 1995. *International trade and regional economics*. Boulder, Colo.: Westview.

Heckscher, E. 1919. The effect of foreign trade on the distribution of income. *Ekonomisk Tidskrift* 21: 497–512.

Heilbroner, R. 1972. *The worldly philosophers*. New York: Simon and Schuster.

———. 1988. *Behind the veil of economics: essays in the worldy philosophy*. New York: W. W. Norton.

Hekman, J. 1980. The product cycle and New England textiles. *Quarterly Journal of Economics* 94: 697–717.

Helper, S. 1991a. Strategy and irreversibility in supplier relations: the case of the U.S. automobile industry. *Business History Review* 65: 781–824.

———. 1991b. How much has really changed between U.S. automakers and their suppliers? *Sloan Management Review* Summer: 15–28.

———. 1994. Three steps forward, two steps back in automotive supplier relations. *Technovation* 14: 633–640.

Henderson, V. 1997. Medium size cities. *Regional Science and Urban Economics* 27: 583–612.

Herrick, B., and Kindleberger, C. 1983. *Economic development*. 4th ed. New York: McGraw-Hill.

Herzog, H., and Schlottmann, A., eds. 1991. *Industry location and public policy*. Knoxville: University of Tennessee Press.

Higgins, B. 1983. From growth poles to systems of interactions in space. *Growth and Change* 14: 3-13.

Higgins, B., and Savoie, D. 1995. *Regional development theories and their application*. New Brunswick, N.J.: Transaction.

———, eds. 1997. *Regional economic development*. Boston: Unwin Hyman.

Hirschman, A. 1958. *The strategy of economic development*. New Haven: Yale University Press.

Holladay, M. 1992. Economic and community development: a southern exposure. Occasional papers of the Kettering Foundation, Dayton, Ohio.

Holmes, J. 1986. The organization and locational structure of production subcontracting. In A. Scott and M. Storper, eds., *Production, work, territory*. Boston: Allen and Unwin.

Hoover, E. 1937. *Location theory and the shoe and leather industries*. Cambridge, Mass.: Harvard University Press.

———. 1971. *An introduction to regional economics*. 1st ed. New York: Knopf.

Hoover, E., and Fisher, J. 1949. *Problems in the study of economic growth*. New York: National Bureau of Economic Research.

Hoover, E., and Giarratani, F. 1984. *An introduction to regional economics*. 3rd ed. New York: Knopf.

Hoover, E., and Vernon, R. 1959. *Anatomy of a metropolis: the changing distribution of people and jobs within the New York metropolitan region*. Cambridge, Mass.: Harvard University Press.

Horiba, Y., and Kirkpatrick, R. 1981. Factor endowments, factor proportions and the allocative efficiency of U.S. interregional trade. *Review of Economics and Statistics* 63: 178-187.

Howarth, R. 1997. Defining sustainability. *Land Economics* 73: 445-447.

Hoyt, H. 1954. Homer Hoyt on the concept of the economic base. *Land Economics* 30: 182-186.

———. 1961. Utility of the economic base method in calculating urban growth. *Land Economics* 37: 51-58.

Hoyt, H., and Weimer, A. 1939. *Principles of urban real estate*. New York: Ronald Press.

Hunt, D. 1989. *Economic theories of development: an analysis of competing paradigms*. Savage, Md.: Barnes and Noble.

Innis, H. 1920. *The fur trade in Canada*. New Haven: Yale University Press.

———. 1933. *Problems of staple production in Canada*. Toronto: University of Toronto Press.

———. 1940. *The cod fishery: the history of an international economy*. New Haven: Yale University Press.

Ioannides, Y. M. 1994. Product differentiation and endogenous growth in a system of cities. *Regional Science and Urban Economics* 24: 461-484.

Isard, W. 1956. *Location and space economy*. Cambridge, Mass.: MIT Press.

Isserman, A. 1977. The location quotient approach to measuring regional economic impacts. *Journal of American Institute of Planners* 43: 33-41.

———. 1980. Estimating export activity in a regional economy: a theoretical and empirical analysis of alternative methods. *International Regional Science Review* 5: 155-184.

———. 1996. It's obvious, it's wrong, anyway they said it years ago? *International Regional Science Review* 19: 27-48.

Jacobs, J. 1969. *The economy of cities*. New York: Random House.

———. 1985. *Cities and the wealth of nations*. New York: Vintage Books.

Jeep, E. 1993. The four pitfalls of local economic development. *Economic Development Quarterly* 7: 237-242.

Jones, H. 1975. *An introduction to modern theories of economic growth*. London: Nelson.

Julien, P., and Raymond, L. 1994. Factors of new technology adoption in the retail sector. *Entrepreneurship Theory and Practice* Summer, 79-90.

Kaldor, N. 1978. The case for regional policies. In *Further essays on economic theory*. London: Gerald Duckworth, 139-154.

Kilby, P. 1988. Breaking the entrepreneurial bottle-neck in developing countries: is there a useful role for government? *Journal of Development Planning* 18: 221-250.

Kirzner, I. 1973. *Competition and entrepreneurship*. Chicago: University of Chicago Press.

———. 1979. *Perception, opportunity, and profit: studies in the theory of entrepreneurship*. Chicago: University of Chicago Press.

———. 1982. The theory of entrepreneurship in economic growth. In D. Sexton, C. Kent, and K. Vesper, eds., *Encyclopedia of Entrepreneurship*. Englewood Cliffs, N.J.: Prentice-Hall, 272-276.

Klier, T. 1994. The impact of lean manufacturing on sourcing relationships. *Economic Perspectives*, Federal Reserve Bank of Chicago, July/August, 8-18.

Knight, F. 1921. *Risk, uncertainty and profit*. Boston, Mass.: Riverside Press.

Kort, J. 1981. Regional economic instability and industrial diversification in the U.S. *Land Economics* 57: 596-608.

Kraybill, D., and Dorfman, J. 1992. A dynamic intersectoral model of regional economic growth. *Journal of Regional Science* 32: 1-17.

Kreinin, M. 1979. *International economics: a policy approach*. New York: Harcourt Brace Jovanovich.

Krikelas, A. 1992. Why regions grow: a review of research on the economic base model. *Economic Review* (Federal Reserve Bank of Atlanta) 77: 16-29.

Krugman, P. 1990. *Rethinking international trade*. Cambridge, Mass.: MIT Press.
———. 1991. *Geography and trade*. Cambridge, Mass.: MIT Press.
———. 1996. *Pop internationalism*. Cambridge, Mass.: MIT Press.
———. 1997. *Development, geography, and economic theory*. Cambridge, Mass.: MIT Press.
Krugman, P., and Obstfeld, M. 1997. *International economics: theory and policy*. Reading, Mass.: Addison-Wesley.
Landreth, H., and Colandeer, D. 1989. *History of economic theory*, 2nd ed. Boston: Houghton Mifflin.
Leff, N. 1979. Entrepreneurship and economic development: the problem revisited. *Journal of Economic Literature* 17: 46-64.
Leibenstein, H. 1968. Entrepreneurship and development. *American Economic Review* 58: 72-83.
———. 1978. *General x-efficiency theory and economic development*. New York: Oxford University Press.
———. 1987. Entrepreneurship, entrepreneurial training, and x-efficiency theory. *Journal of Economic Behavior* 8: 191-205.
Leontief, W. 1953. Domestic production and foreign trade; the American capital position re-examined. *Proceedings of the American Philosophical Society* 97: 332-49.
———. 1956. Factor proportions and the structure of American trade: further theoretical and empirical analysis. *Review of Economics and Statistics* 38: 386-407.
LeSage, J. 1990. Forecasting metropolitan employment using an export-base error-correction model. *Journal of Regional Science* 30: 307-324.
LeSage, J., and Reed, J. 1989. The dynamic relationship between export, local, and total area employment. *Regional Science and Urban Economics* 19: 615-636.
Leven, C. 1964. *Potential applications of systems analysis to regional economic planning*. Pittsburgh: Center for Regional Economic Studies, University of Pittsburgh.
Levy, J. 1990. What local economic developers actually do. *Journal of the American Planning Association* 56: 153-160.
———. 1992. The U.S. experience with local economic development. *Environment and Planning C* 10: 51-60.
Lewis, W., and Prescott, J. 1972. Urban-regional development and growth centers: an econometric study. *Journal of Regional Science* 12: 57-70.
Little, J., and Triest, R. 1996. Technology diffusion in U.S. manufacturing: the geographic dimension. In J. Fuhrer and J. Little, eds., *Technology and Growth*, 215-259, conference proceedings of the Federal Reserve Bank of Boston. Boston, Mass.: Federal Reserve Bank of Boston.

Lösch, A. 1954. *The economics of location*. 2nd rev. ed. W. Woglom, trans. New Haven: Yale University Press.

Lucas, R., Jr. 1988. On the mechanics of economic development. *Journal of Monetary Economics* 22: 3-42.

Lumpkin, J., and Ireland, R. 1988. Screening practices of new business incubators: the evaluation of critical success factors. *American Journal of Small Business* Spring, 59-81.

Malecki, E. 1981. Product cycles, innovation cycles, and regional economic change. *Technological Forecasting and Social Change* 19: 291-306.

———. 1989. Technology, employment, and regional competitiveness. *Economic Development Quarterly* 3: 331-338.

———. 1997. *Technology and economic development: the dynamics of local, regional and national competitiveness*. 2nd ed. Essex, England: Addison Wesley Longman.

Malizia, E. 1985. *Local economic development: a guide to practice*. New York: Praeger.

———. 1986. Economic development in smaller cities and rural areas. *Journal of the American Planning Association* 52: 489-499.

———. 1987. Discounting employment as a measure of economic development benefits. *Economic Development Quarterly* 1: 374-378.

———. 1990. Economic growth and economic development: concepts and measures. *Review of Regional Studies* 20: 30-36.

———. 1994. A redefinition of economic development. *Economic Development Review* 12: 83-84.

———. 1996. Two strategic paths to competitiveness. *Economic Development Review* 14: 7-9.

Malizia, E., and Feser, E. 1994. The proper niche of modern portfolio theory in regional analysis. Working Paper S95-10: Center for Urban and Regional Studies, University of North Carolina at Chapel Hill.

Malizia, E., and Reid, D. 1976. Perspectives and strategies for U.S. regional development. *Growth and Change* 7: 41-47.

Markusen, A. 1985. *Profit cycles, oligopoly, and regional development*. Cambridge, Mass.: MIT Press.

———. 1986. Defense spending and the geography of high-tech industries. In J. Rees, ed., *Technology, regions, and policy*. Syracuse: Rowman and Littlefield, 94-119.

———. 1996. Sticky places in slippery space: a typology of industrial districts. *Economic Geography* 72: 293-313.

Marshall, A. 1961 [1890]. *Principles of economics: an introductory volume*. 9th (Variorum) ed. London: Macmillan.

Martin, H. 1994. There is a "magic" in Tupelo-Lee County, Northeast Mississippi. *Economic Development Review* 12: 78-82.

Martin, R. 1990. Flexible futures and post-Fordist places. *Environment and Planning A* 22: 1276-1280.

Martin, R., and Sunley, P. 1996. Paul Krugman's geographical economics and its implications for regional development theory: a critical assessment. *Economic Geography* 72: 259-292.

Marx, K. 1967. In F. Engels, ed., *Capital*. 3 vols. New York: International Publishers.

Mathur, V., and Rosen, H. 1972. An econometric export base model of regional growth: a departure from conventional techniques. In Wilson, A. ed., *Patterns and processes in urban and regional systems, London Papers in Regional Science 3*, 31-43. London: Pion.

———. 1974. Regional employment multiplier: a new approach. *Land Economics* 50: 93-96.

———. 1975. Regional employment multiplier: a new approach: reply. *Land Economics* 51: 294-295.

McCann, P., and Fingleton, B. 1996. The regional agglomeration impact of just-in-time input linkages: evidence from the Scottish electronics industry. *Journal of Scottish Political Economy* 43, 5: 493-518.

McClelland, D. 1961. *The achieving society*. Princeton, N.J.: Van Nostrand.

McCombie, J. 1988a. A synoptic view of regional growth and unemployment: I—The neoclassical theory. *Urban Studies* 25: 267-281.

———. 1988b. A synoptic view of regional growth and unemployment: II—The post-Keynesian theory. *Urban Studies* 25: 399-417.

McCrone, G. 1969. *Regional policy in Britain*. London: Allen and Unwin.

Meier, G. 1984. *Leading issues in economic development*. 4th ed. New York: Oxford University Press.

Meier, G., and Baldwin, R. 1957. *Economic development: theory, history, policy*. New York: John Wiley and Sons.

Miernyk, W. 1977. The changing structure of the Southern economy. In B. Liner and L. Lynch, eds., *The economics of Southern growth*. Washington: Economic Development Administration, Office of Economic Research.

Miller, M.; Gibson, L.; and Wright, N. 1991. Location quotient: a basic tool for economic development. *Economic Development Review* 9: 65-68.

Moore, B. 1966. *Social origins of dictatorship and democracy: lord and peasant in the making of the modern world*. Boston: Beacon Press.

Moore, C. 1975. A new look at the minimum requirements approach to regional economic analysis. *Economic Geography* 51: 350-356.

Morfessis, I. 1994. A cluster-analytic approach to identifying and developing state target industries: the case of Arizona. *Economic Development Review* 12: 33-7.

Moriarty, B. 1980. *Industrial location and community development*. Chapel Hill: University of North Carolina Press.

———. 1983. Hierarchies of cities and the spatial filtering of industrial development. *Papers of the Regional Science Association* 53: 54–82.

———. 1992. The manufacturing employment longitudinal density distribution in the USA. *Review of Regional Studies* 22: 1–23.

Moroney, J. 1972. *The structure of production in American manufacturing*. Chapel Hill: University of North Carolina Press.

Moroney, J., and Walker, J. 1966. A regional test of the Heckscher-Ohlin hypothesis. *Journal of Political Economy* 74: 573–586.

Moseley, M. 1974. *Growth centres in spatial planning*. New York: Pergamon Press.

Moss Kanter, R. 1995. *World class: thriving locally in the global economy*. New York: Simon and Schuster.

Myrdal, G. 1944. *An American dilemma, the Negro problem and modern democracy*. New York: Harper.

———. 1957. *Economic theory and underdeveloped regions*. New York: Harper and Row.

NC ACTS. 1996. Results of the North Carolina competitiveness survey. Research Triangle Park: North Carolina Alliance for Competitive Technologies.

Nelson, R. 1956. A theory of the low-level equilibrium trap in underdeveloped economies. *American Economic Review* 46: 894–908.

Noponen, H.; Graham J.; and Markusen, A., eds. 1993. *Trading industries, trading regions: international trade, American industry, and regional economic development*. New York: Guilford Press.

Norcliffe, G. 1983. Using location quotients to estimate the economic base and trade flows. *Regional Studies* 17: 161–168.

North, D. 1955. Location theory and regional economic growth. *Journal of Political Economy* 63: 243–258.

Norton, R., and Rees, J. 1979. The product cycle and the spatial decentralization of American manufacturing. *Regional Studies* 13: 141–151.

Noyelle, T. 1983. The rise of advanced services: some implications for economic development in U.S. cities. *Journal of the American Planning Association* 49: 280–290.

Noyelle, T., and Stanback, T. 1984. *The economic transformation of American cities*. Totowa, N.J.: Rowman and Allanheld.

Nurske, R. 1953. *Problems of capital formation in underdeveloped countries*. Oxford: Basil Blackwell.

Ohlin, B. 1933 [1967]. *Interregional and international trade*. Rev. ed. Cambridge, Mass.: Harvard University Press.

Ohmae, K. 1995. *The end of the nation state: the rise of regional economies*. New York: Free Press.

Ollman, B. 1971. *Alienation: Marx's conception of man in a capitalist society*. Cambridge: Cambridge University Press.

Osborne, D. 1988. *Laboratories of democracy*. Boston: Harvard Business School Press.

Palivos, T., and Wang, P. 1996. Spatial agglomeration and endogenous growth. *Regional Science and Urban Economics* 26: 645-669.

Palm, R. 1981. *The geography of American cities*. New York: Oxford University Press.

Park, S. O. 1997. Dynamics of new industrial districts and regional economic development. Paper presented at the 1997 International Symposium on Industrial Park Development and Management, Taipei, Taiwan.

Parr, J. 1970. Models of city size in an urban system. *Papers of the Regional Science Association* 25: 221-253.

———. 1973. Growth poles, regional development, and central place theory. *Papers of the Regional Science Association* 31: 173-212.

Perloff, H.; Dunn, E.; Lampard, E.; and Muth, R. 1960. *Regions, resources and economic growth*. Lincoln: University of Nebraska Press.

Perloff, H., and Wingo, H., Jr. 1968. *Issues in urban economics*. Baltimore: Johns Hopkins University Press.

Perroux, F. 1950a. Economic space: theory and applications. *Quarterly Journal of Economics* 64: 89-104.

———. 1950b. The domination effect and modern economic theory. *Social Research* 17: 188-206.

———. 1988. The role of development's new place in a general theory of economic activity. In *Regional economic development: essays in honor of François Perroux*. Sydney: Unwin Hyman.

Pinchot, G. 1985. *Intrapreneuring: why you don't have to leave the corporation to become an entrepreneur*. New York: Harper and Row.

Piore, M., and Sabel, C. 1984. *The second industrial divide*. New York: Basic Books.

Polanyi, K. 1944. *The great transformation*. New York and Toronto: Farrar and Rinehart.

———. 1968. The self-regulating market and the fictitious commodities: land, labor, and money (1944) In G. Dalton, ed., *Primitive, archaic and modern economies*. Boston: Beacon Press.

Pollert, A. 1988. Dismantling flexibility. *Capital and Class* 34: 42-75.

Popovich, M., and Buss, T. 1990. 101 ideas for stimulating rural entrepreneurship and new business development. *Economic Development Review* 8: 26-32.

Porter, M. 1990. *The competitive advantage of nations*. New York: Free Press.

———. 1997. New strategies for inner-city economic development. *Economic Development Quarterly* 11: 11-27.

Pratt, R. 1968. An appraisal of the minimum requirements technique. *Economic Geography* 44: 117-124.

Pred, A. 1976. The interurban transmission of growth in advanced economies: empirical findings versus regional planning assumptions. *Regional Studies* 10: 151-171.

Rabellotti, R. 1995. Is there an "industrial district model"? Footwear districts in Italy and Mexico compared. *World Development* 23: 29-41.

Ranis, G., and Schultz, T., eds. 1988. *The state of development economics: progress and perspectives*. Oxford: Blackwell.

Rephann, T., and Shapira, P. 1994. Survey of technology use in West Virginia manufacturing. Research Paper 9401, West Virginia University, Regional Research Institute.

Ricardo, D. 1912 [1817]. *The principles of political economy and taxation*. New York: E. P. Dutton.

Richardson, H. 1973. *Regional growth theory*. London: Macmillan.

———. 1978. *Regional economics*. Urbana: University of Illinois Press.

———. 1985. Input-output and economic base multipliers: looking backward and forward. *Journal of Regional Science* 25: 607-661.

Rima, I. 1991. *Development of economic analysis*. 5th ed. Homewood, Ill.: Richard D. Irwin.

Robinson, E. 1931. *The structure of competitive industry*. Cambridge: Cambridge University Press.

Robinson, G., and Salih, K. 1971. The spread of development around Kuala Lumpur: a methodology for an exploratory test of some assumptions of the growth-pole model. *Regional Studies* 5: 303-314.

Robinson, J., and Eatwell, J. 1973. *An introduction to modern economics*. London: McGraw Hill.

Roelandt, T.; den Hertog, P.; van Sinderen, J.; and Vollaard, B. 1997. Cluster analysis and cluster policy in the Netherlands. Paper presented at the OECD Workshop on Cluster Analysis and Cluster Policies, Amsterdam, Netherlands, October 9-10, 1997.

Romer, P. 1986. Increasing returns and long-run growth. *Journal of Political Economy* 94: 1002-37.

Rondinelli, D. 1983. *Secondary cities in developing countries: policies for diffusing urbanization*. Beverly Hills: Sage Publications.

Rosenfeld, S. 1992. *Competitive manufacturing*. New Brunswick, N.J.: Center for Urban Policy Research.

———. 1995. *Industrial-strength strategies: regional business clusters and public policy*. Washington, D.C.: Aspen Institute.

Rosenstein-Rodan, P. 1943. Problems of industrialization of eastern and south-eastern Europe. *Economic Journal* 53: 202-211.

Rostow, W. 1964. *View from the seventh floor*. New York: Harper and Row.

Rubin, H. 1988. Shoot anything that flies; claim anything that falls: conversations with economic development practitioners. *Economic Development Quarterly* 2: 236-251.

Russell, B. 1945. *A history of western philosophy*. New York: Simon and Schuster.

Salter, W. 1966. *Productivity and technical change*. Cambridge: Cambridge University Press.

Samuelson, P. 1948. International trade and the equalization of factor prices. *Economic Journal*. 58: 163-184.

Saxenian, A. 1994. *Regional advantage: culture and competition in Silicon Valley and Route 128*. Cambridge, Mass.: Harvard University Press.

Sayer, A. 1990. Post-Fordism in question. *International Journal of Urban and Regional Research* 13: 666-695.

Schoenberger, E. 1988. From Fordism to flexible accumulation: technology, competitive strategies, and international location. *Environment and Planning D: Society and Space* 6: 245-262.

Schumacher, E. 1973. *Small is beautiful*. New York: Harper and Row.

Schumpeter, J. 1934. *The theory of economic development*. Cambridge, Mass.: Harvard University Press.

———. 1947. The creative response in economic history. *Journal of Economic History* 7: 149-159.

———. [1950]. *Capitalism, socialism, and democracy*. New York: Harper.

Schweke, W. 1985. Why local governments need an entrepreneurial policy. *Public Management* 67: 3-6.

Scitovsky, T. 1954. Two concepts of external economies. *Journal of Political Economy* 62: 70-82.

Scott, A. 1986. Industrial organization and location: division of labor, the firm, and spatial process. *Economic Geography* 63: 215-31.

———. 1988. *Metropolis: from the division of labor to urban form*. Berkeley: University of California Press.

———. 1992. The collective order of flexible production agglomerations: lessons for local economic development policy and strategic choice. *Economic Geography* 68: 219-33.

Scott, A., and Storper, M. 1987. High technology industry and regional development: a theoretical critique and reconstruction. *International Social Science Journal* 39: 215-232.

Seers, D. 1969. The meaning of development. In C. Wilber, ed., *The political economy of development and underdevelopment*. New York: Random House.

———. 1970 [1963]. The stages of economic growth of primary producers in the middle of the twentieth century. In R. Rhodes, ed., *Imperialism and underdevelopment*. New York: Monthly Review Press.

Sexton, D. L., and Kasarda, J. D. 1992. *The state of the art of entrepreneurship*. Boston: PWS-Kent.

Shapero, A. 1977. *The role of entrepreneurship in economic development at the less-than-national level*. Washington: Economic Development Administration, U.S. Department of Commerce, January.

———. 1981. Entrepreneurship: key to self renewing economies. *Economic Development Commentary* 7: 19-23.

Smith, Adam. 1976 [1759]. *The theory of moral sentiments*. Indianapolis: Liberty Classics.

Smith, Adam. 1937 [1776]. *The wealth of nations*. E. Cannan, ed. New York: Modern Library.

Smith, T., and Ferguson, D. 1995. Targets of state and local economic development policy. *Regional Economic Digest* 6: 4-11.

Smith, T., and Fox, W. 1991. Economic development programs for states in the 1990s. *Regional Economic Development and Public Policy*. Kansas City: Federal Reserve Bank of Kansas City.

Solow, R. 1956. A contribution to the theory of economic growth. *Quarterly Journal of Economics* 70: 65-94.

Spiegel, H. 1983. *The growth of economic thought*. Durham, N.C.: Duke University Press.

Stabler, J., and St. Louis, L. 1990. Embodied inputs and the classification of basic and nonbasic activity: implications for economic base and regional growth analysis. *Environment and Planning A* 22: 1667-1676.

Stebbins, D. 1995. Local economic development planning and practice case study: Buffalo, NY. Master's Project paper. Department of City and Regional Planning, University of North Carolina-Chapel Hill.

Steiner, P. 1987. Contrasts in regional potentials: some aspects of regional economic development. *Papers of the Regional Science Association* 61: 79-92.

Stern, N. 1989. The economics of development: a survey. *Economic Journal* 99: 598-685.

Sternberg, E. 1993. Justifying public intervention without market externalities: Karl Polanyi's theory of planning in capitalism. *Public Administration Review* 53: 100-109.

Sternberg, E. 1987. A practitioner's classification of economic development policy instruments, with some inspiration from political economy. *Economic Development Quarterly* 1: 149-161.

Stigler, G. 1951. The division of labour is limited by the extent of the market. *Journal of Political Economy* 59: 185-193.

Stough, R. 1998. Endogenous growth in a regional context: Introduction. *Annals of Regional Science* 32, 1: 1-5.

Strachan, H. 1976. *Family and other business groups in economic development: the case of Nicaragua*. New York: Praeger.

Streeten, P. 1959. Unbalanced growth. *Oxford Economic Papers* 11: 167-190.

Suarez-Villa, L. 1988. Metropolitan evolution, sectoral economic change, and the city-size distribution. *Urban Studies* 25: 1-20.

Swamidass, P. 1994. *Technology on the factory floor II: benchmarking manufacturing technology use in the United States*. Washington, D.C.: Manufacturing Institute.

Swamy, D. 1967. Statistical evidence of balanced and unbalanced growth. *Review of Economics and Statistics* 49: 288-303.

Sweeney, S., and Feser, E. 1998. Plant size and clustering of manufacturing activity. *Geographical Analysis* 30, 1: 45-64.

Taylor, M. 1986. The product-cycle model: a critique. *Environment and Planning A* 18: 751-761.

Teitz, M. 1994. Changes in economic development theory and practice. *International Regional Science Review* 16: 101-106.

Tellier, L. 1997. A challenge for regional science: revealing and explaining the global spatial logic of economic development. *Papers in Regional Science* 76: 371-384.

Thirlwall, A. 1979. The balance of payments constraint as an explanation of international growth rate differences. *Banca Nazionale Lavoro Quarterly Review* 128: 45-53.

———. 1980. Regional problems are "balance-of-payments" problems. *Regional Studies* 14: 419-425.

Thomas, M. 1975. Growth pole theory, technological change, and regional economic growth. *Papers of the Regional Science Association* 34: 3-25.

Thompson, W. 1965. *A preface to urban economics*. Baltimore: Johns Hopkins University Press.

———. 1968. Internal and external factors in the development of urban economics. In Perloff, H., and Wingo, L. eds., *Issues in urban economics*. Baltimore: Johns Hopkins Press, 43-62.

Thompson, W., and Thompson, P. 1985. From industries to occupations: rethinking local economic development. *Commentary* 9: 12-18.

———. 1987. National industries and local occupational strengths. *Urban Studies* 24: 547-560.

Thwaites, A., and Oakey, R., eds. 1985. *The regional economic impact of technical change*. London: Frances Pinter.

Tiebout, C. 1956a. Exports and regional economic growth. *Journal of Political Economy* 64: 160-164. Reprinted in Friedmann and Alonso (1964, pp. 256-260).

———. 1956b. Exports and regional economic growth, a rejoinder. *Journal of Political Economy* 64: 169. Reprinted in Friedmann and Alonso (1964, p. 265).

———. 1962. *The community economic base study*. New York: Committee for Economic Development.

Toye, J. 1987. *Dilemmas of development*. London: Basil Blackwell.

Tremblay, G. 1993. Moving towards a value-added society: Quebec's new economic development strategy. *Economic Development Review* 11: 18–20.

Ullman, E., and Dacey, M. 1960. The minimum requirements approach to the urban economic base. *Papers of the Regional Science Association* 6: 174–194.

Ullman, E.; Dacey, M.; and Brodsky, H. 1969. *The economic base of American cities*. Seattle: University of Washington Press.

United States President. 1997. *Economic report of the President*. Washington, D.C.: Government Printing Office.

U.S. Department of Commerce. 1989. *Manufacturing technology, 1988*. Washington, D.C.

U.S. Department of Commerce. 1993. *Manufacturing technology: factors affecting adoption, 1991*. Washington, D.C.

Vapnarsky, C. 1969. On rank size distributions of cities. *Economic Development and Cultural Change* 17: 584–595.

Vernon, R. 1963. *Metropolis 1985*. Garden City, New York: Doubleday.

———. 1966. International investment and international trade in the product cycle. *Quarterly Journal of Economics* 80: 190–207.

———. 1979. The product cycle hypothesis in a new international environment. *Oxford Bulletin of Economics and Statistics* 41: 255–267.

Viner, J. 1931. Cost curves and supply curves. *Zeitschrift für National... konomie* 3: 23–46.

Waitt, G. 1993. Say bye to Hyundai and hi to Korean autoparts? Restructuring in the Korean automobile industry in the 1990s. *Tijdschrift voor Economische en Sociale Geografie* 84, 3: 198–206.

Wasylenko, M. 1991. Empirical evidence on interregional business location decisions and the role of fiscal incentives in economic development. In H. Herzog and A. Schlottmann, eds., *Industry Location and Public Policy*. Knoxville: University of Tennessee Press.

Weber, A. 1929. *Theory of the location of industries*. C. Friedrich, trans. Chicago: University of Chicago Press.

Weinstein, B., and Firestine, R. 1978. *Regional growth and decline in the United States: the rise of the sunbelt and the decline of the northeast*. New York: Praeger.

Williamson, O. 1975. *Markets and hierarchies*. New York: Free Press.

———. 1985. *The economic institutions of capitalism*. New York: Free Press.

Wilson, P. 1990. The new Maquiladoras: flexible production in low-wage regions. In K. Fatemi, ed., *The Maquiladora industry: economic solution or problem*. New York: Praeger.

Wozniak, G. 1987. Human capital, information, and the early adoption of new technology. *Journal of Human Resources* 22: 101-112.
Yarzebinski, J. 1992. Understanding and encouraging the entrepreneur. *Economic Development Review* 10: 32-35.
Young, A. 1928. Increasing returns and economic progress. *Economic Journal* 38: 527-542.
Youtie, J., and Shapira, P. 1995. *Manufacturing needs, practices, and performance in Georgia: 1994 Georgia manufacturing technology survey*. Georgia Manufacturing Extension Alliance, Georgia Tech Economic Development Institute.
Youtopolous, P., and Lau, L. 1970. A test for balanced and unbalanced growth. *Review of Economics and Statistics* 52: 376-384.
Zweig, M. 1973. Foreign investment and the aggregate balance of payments for underdeveloped countries. *Review of Radical Political Economies* 5: 13-18.

Suggestions for Further Reading

Aitken, H., ed. 1965. *Explorations in enterprise*. Cambridge, Mass.: Harvard University Press.
Allen, D., and Hayward, D. 1990. The role of new venture formation/entrepreneurship in regional economic development: a review. *Economic Development Quarterly* 4: 55-63.
Allen, D.; Bazan, E.; and Hendrickson-Smith, J. 1986. Gritty and flashy entrepreneurs in the Atlanta metropolitan area. In R. Ronstadt et al., eds., *Frontiers of entrepreneurship research: proceedings of the 1982 entrepreneurship research conference*. Wellesley, Mass.: Center for Entrepreneurial Studies, Babson College, 303-312.
Christian, C., and Harper, R., eds. 1982. *Modern metropolitan systems*. Columbus, Ohio: Charles E. Merrill.
Friedman, R. 1987. The role of entrepreneurship in rural development. In B. Honadle and J. Reid, eds., *National rural entrepreneurship symposium* 1-6. SRDC series no. 97, Southern Rural Development Center.
Heckscher, E. 1935. *Mercantilism*. London: Allen and Unwin.
Higgins, B. 1978. Development poles: do they exist? In F. Lo and K. Salih, eds., *Growth pole strategy and regional development policy*, 229-242. New York: Pergamon Press.
Hoselitz, B. 1960. *Theories of economic growth*. Glencoe, Ill.: Free Press.
Isard, W. 1954. Location theory and trade theory: short-run analysis. *Quarterly Journal of Economics* 68: 305-320.
Kaldor, N. 1980a. *Essays on economic stability and growth*. 2nd ed. New York: Holmes and Meier.

———. 1980b. *Essays on value and distribution*. 2nd ed. New York: Holmes and Meier.

———. 1985. *Economics without equilibrium*. Armonk, N.Y.: M.E. Sharpe.

Kilby, P. 1971. *Entrepreneurship and economic development*. New York: Free Press.

Malecki, E. 1983. Technology and regional development: a survey. *International Regional Science Review* 8: 89–125.

Malizia, E., and Rubin, S. 1982. New strategies for rural economic development. *Carolina Planning* 8: 39–46.

Markusen, A.; Hall, P.; and Glasmeier, A. 1986. *High tech America*. Boston: Allen and Unwin.

Moroney, J. 1975. Natural resource endowments and comparative advantage. *Journal of Regional Science* 15: 139–150.

Moroney, J. 1970. Factor prices, factor proportions, and regional factor endowments. *Journal of Political Economy* 78: 158–164.

Moroney, J. 1967. The strong-factor intensity hypothesis: a multi-sectoral test. *Journal of Political Economy* 75: 241–249.

Pfouts, R., ed. 1960. *The techniques of urban economic analysis*. Trenton, N.J.: Chandler-Davis Publishing.

Rostow, W. 1960. *The stages of growth, a non-communist manifesto*. Cambridge: Cambridge University Press.

Tiebout, C. 1956. The urban economic base reconsidered. *Land Economics* 22: 17–23.

Wilber, C. 1979. *The political economy of development and underdevelopment*. New York: Random House.

Professional and Scholarly Economic Development Journals

Economic Development Commentary. Washington, D.C.: National Council for Urban Economic Development. Subjects: economic development; urban economics; community development. (Former title: *Commentary*.)

Research on Economic Inequality. "A research annual." Greenwich, Conn.: JAI Press. Subjects: income distribution; economic development.

Economic Development Abroad. Washington, D.C.: National Council for Urban Economic Development, International Center, Academy for State and Local Government. Subjects: economic development; community development.

CUED Economic Developments / National Council for Urban Economic Development. Washington, D.C.: National Council for Urban Economic

Development. Subjects: urban economics; community development; urban policy. (Former title: *CUED Urban Economic Developments.*) Current frequency: two issues per month.

Advances in the Study of Entrepreneurship, Innovation, and Economic Growth. Greenwich, Conn.: JAI Press. Subjects: Entrepreneurship; new business enterprises; technological innovations; venture capital; economic development.

Research in Human Capital and Development. Greenwich, Conn.: JAI Press. Subjects: economic development; human capital. Except for Volume 1, each volume has a distinctive title. Includes bibliographies.

Economic Development Review. Schiller Park, Ill.: American Economic Development Council in cooperation with the Southern Industrial Development Council and other sponsors. Subjects: economic development; U.S. economic conditions.

Community Development Digest. Washington, D.C.: Community Development Clearinghouse. Subjects: community development; housing law and legislation. (Former title: *Housing and Renewal Index.*) Current frequency: two issues monthly.

Economic Development Quarterly. "The journal of American economic revitalization." Newbury Park, Calif.: Sage Publications. Subjects: economic development; economics; U.S. economic conditions.

Environment and Planning, various editions. C: Government and Policy. London: Pion Ltd. Subjects: policy sciences; public administration. (Former title: *Environment and Planning*, 1969–1973.)

Growth and Change. "A journal of regional development." Lexington, University of Kentucky, College of Business and Economics. Subjects: regional planning.

International Journal of Urban and Regional Research. London: E. Arnold. Subjects: city planning; regional planning. Current frequency: quarterly.

International Regional Science Review. Philadelphia: Regional Science Association. Subjects: environmental policy; regional planning. Current frequency: two issues per year.

Journal of Regional Science. Amherst, Mass.: Regional Science Research Institute. Subjects: economics; regional planning. Includes bibliographies. Current frequency: quarterly.

Journal of Urban Affairs. Blacksburg, Va.: Division of Environment and Urban Systems, Virginia Polytechnic Institute and State University. Subjects: cities and towns; city planning; urban policy. Current frequency: quarterly.

Regional Studies (Journal of the Regional Studies Association). Cambridge: Cambridge University Press. Subjects: regional planning. Current frequency: bimonthly.

Regional Science and Urban Economics. Amsterdam: North-Holland. Subjects: regional planning; urban economics; economics. (Former title: *Regional and Urban Economics*).

The Review of Regional Studies. Blacksburg, Va.: Southern Regional Sciences Association. Subject: regional planning. Current frequency: three issues per year. Population index.

Rural Development Perspectives. Washington, D.C.: United States Department of Agriculture, Economics, Statistics, and Cooperatives Service. Subject: rural development.

Urban Affairs Quarterly. New York: Sage Publications. Subjects: city planning; metropolitan areas. (Later title: *Urban Affairs Review*, Thousand Oaks, Calif.).

Urban Studies. Edinburgh: Longman Group Ltd. Subjects: cities and towns; city planning.

Electronic Sources

Agencies and Federal Corporations

Appalachian Regional Commission
www.arc.gov/

Tennessee Valley Authority
www.tva.com/

U.S. Department of Housing and Urban Development
www.hud.gov/

U.S. Economic Development Administration
www.doc.gov/eda/

U.S. Small Business Administration
www.sbaonline.sba.gov/

Informational Sites

Rutgers University Economic Development Planning Webpage
crab.rutgers.edu/~raetsch/edplan.html

Hubert H. Humphrey Institute of Public Affairs
Economic Development Web Site
www.hhh.umn.edu/Centers/SLP/edweb/

Center for Community Economic Development, University of Wisconsin
www.uwex.edu/ces/cced/
ElectriCities of North Carolina, Nationwide Economic Development Web Sites
www.electricities.com/servlis1.html

Organizations

American Economic Development Council
www.aedc.org/
National Association of Development Organizations
www.nado.org/index/html
Corporation for Enterprise Development
www.cfed.org/
Council for Urban Economic Development
www.cued.org/cued/index.html
U.S. Conference of Mayors
www.usmayors.org/uscm/home.html

Institutes and Research Houses

Lincoln Institute of Land Policy
www.lincolninst.edu/main.html
The Urban Institute
www.urban.org/
Urban Land Institute
www.uli.org/
W. E. Upjohn Institute for Employment Research
www.upjohninst.org/

Index

Note: Page numbers in italics refer to tables.

D

E

F

G

T

U